ISBN 978-1-332-43791-7
PIBN 10259866

English
Français
Deutsche
Italiano
Español
Português

www.forgottenbooks.com

Mythology Photography **Fiction**
Fishing Christianity **Art** Cooking
Essays Buddhism Freemasonry
Medicine **Biology** Music **Ancient**
Egypt Evolution Carpentry Physics
Dance Geology **Mathematics** Fitness
Shakespeare **Folklore** Yoga Marketing
Confidence Immortality Biographies
Poetry **Psychology** Witchcraft
Electronics Chemistry History **Law**
Accounting **Philosophy** Anthropology
Alchemy Drama Quantum Mechanics
Atheism Sexual Health **Ancient History**
Entrepreneurship Languages Sport
Paleontology Needlework Islam
Metaphysics Investment Archaeology
Parenting Statistics Criminology
Motivational

ROAD ALONG THE LAKE OF THUN, SWITZERLAND.

MERICAN HIGHWAYS

A POPULAR ACCOUNT OF THEIR CON-DITIONS AND OF THE MEANS BY WHICH THEY MAY BE BETTERED

BY

N. S. SHALER

DEAN OF THE LAWRENCE SCIENTIFIC SCHOOL OF HARVARD UNIVERSITY

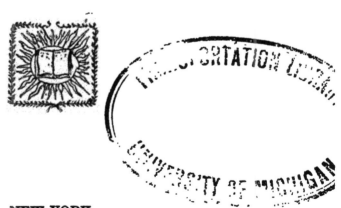

NEW YORK

THE CENTURY CO.

1896

THE DE VINNE PRESS.

LIST OF ILLUSTRATIONS

CONTENTS

CHAPTER X

CHAPTER XI

CHAPTER XII

CHAPTER XIII

PREFACE

THE historian of this country for the century which is now drawing to its close is likely to note the fact that the people of the United States bore in a singularly patient manner with the evils arising from poor carriage roads until near the end of the tenth decade, and that they then were suddenly aroused to a sense of the sore tax the ill condition of these necessary features of civilization had long inflicted upon them. Let us hope that he may be able to say that in approaching this great economic problem, they did so in a manner which showed that they were well informed as to the conditions under which they could deal with it in the light of the previous experience of men, and with the help which the resources of modern science could afford them.

There is always danger, as in any popular uprising directed against ancient and long-endured evils, that the people may, even if the matters are of a purely economic nature, act hastily under the guidance of enthusiasts and with little regard to the help which learning may give them. The result of such action is necessarily a considerable waste of capital, and in almost all cases it leads to much discouragement, and thus to the loss of the spirit which, properly guided, may bring a large accomplish-

ment. An instance of this kind is known to those who
have attentively followed the history of transportation in
this country. In the early part of the nineteenth century,
before the time of railways, there was a curious enthusiasm for canal-building. Guided by the profit that they
had had from the admirable natural waterways of the
country, the people sought to make artificial paths for
boats in various directions, from one great stream to another, and from the Mississippi to the Atlantic. In their
enthusiasm they undertook many enterprises which under
no conditions could have proved profitable. All these
ways intended for distant transportation, except the Erie
Canal, became failures, with the result that, except in such
local work as that which gave access to Lake Superior or
afforded a passage around the falls of the Ohio, canals
came to be regarded with contempt. It was left to the
people of other countries to extend the use of canals and
to show that, even in competition with railways, this
method of carriage might have a great value.

Those who have the betterment of our American highways at heart should do all in their power to guide, direct,
and even restrain the present movement toward their improvement, so that enthusiasm may be guided by a business sense, to the end that we may attain a system of
ways properly related to the needs of the country. It is
with this view that the writer has undertaken to prepare
the following chapters on problems afforded by American
roads. They are not intended to constitute a complete
treatise on road-making. A work of that nature must be
framed on different lines from this; it must be addressed
to professional engineers. Works of this sort already
abound. Many of them are excellent. None of them,
however, so far as is known to the writer, are fitted to

serve as guides for those who wish to understand the general aspects of the highway problem, or who would learn what kind of road may be contrived to meet the needs of the varied surroundings, natural and artificial, in which our people find themselves.[1] They tell in an excellently detailed way how high-grade roads should be built and repaired, but in most cases their proposed constructions are exceedingly costly, even where the materials of which they are to be made are readily accessible. In great areas of country such as are found in the Mississippi valley, where for scores if not hundreds of miles in any direction one may seek in vain for rocks fit to be used as broken stone in forming Macadam roads, these admirable works serve only to dishearten those who would better their ways.

Those who would help the cause of better roads in this country should approach the problem in a large way. They should, in the first place, obtain a general acquaintance as to the influence which roads, as a means of intercourse, exert on the advance of civilization. Next, they should understand the history of the development of such roads, a history which is curiously linked with that of civilization, and is therefore full of interest quite apart from its economic application. It is important also that all who wish to promote the cause should gain a clear sense as to the relation of roads to the topographic, geological, and climatal conditions of the country. Because of the neglect of these features there is great danger that

---

[1] My colleague, Mr. W. E. McClintock, of the Massachusetts Highway Commission, has in preparation a detailed work on road-building from the point of view of the engineer. This work will give the results of many years' professional practice in this field of construction.

the effort to import the methods and apply the experience of foreign countries, or even of districts in this land, to any particular field will lead to grave blunders. In a word, the student, if he would be helpful to himself and others in this matter, must be prepared to consider any road as an extremely local problem, in dealing with which continual reference must be had to all the physical conditions of the ground it traverses, as well as to the probable future of the population in the area which it is to serve. It is because the existing treatises neglect these important considerations that this book has been prepared.

It should be clearly understood that the main object of this book is to consider the conditions of rural ways and not those of city streets. This latter group of roads presents a peculiar class of difficulties which have, in many cases, to be met by devices quite other than those which may be made to serve in country districts. Therefore, while mere incidental reference will be made to these urban problems, questions as to the construction and management of city streets may best be left to the engineers who have them in charge.

It should also be said that certain details of construction and maintenance which are adopted in the roads of the Old World are not considered in this work, for the reason that they do not seem to be adapted to the conditions which exist in this country. The aim of the writer is, in a word, to provide a general account of roads which may fit the needs of the American public.

It seems fit that he should set forth in brief the experience which has made it proper for him to undertake the work which is here set forth. This in outline is as follows: During the Civil War he was in a position to learn the meaning of wheelways in the critical work of campaigns; since that time he has been much interested in road-making and in geological work carried on in different parts of this country and Europe, and has paid close attention to the physical conditions of such ways. For more than four years he has been engaged as one of the three members of the Massachusetts Highway Commission in studying the conditions of the roads in that commonwealth, and developing a plan for their betterment by State constructions. In connection with his colleagues he has had to do with the laying out and constructing of about one hundred roads, and has been particularly engaged in a study of the relations of the road-building materials to the needs of these ways. He has had, moreover, to do with planning, in the Scientific School of Harvard University, a system for the instruction of engineers in road-making, a course which is carried on under the direction of William E. McClintock, one of the members of the above-mentioned Highway Commission. He has prepared two reports for the United States Geological Survey, one entitled "The Geology of Highway Materials," and the other "The Road-building Materials of Massachusetts."

Those who desire more detailed statements concerning the methods of constructing and maintaining particular kinds of roads than are given in the following chapters, may advantageously consult the list of works which is given in the Appendix. This list contains, with few exceptions, only works in English, and includes those only which have some value to the non-professional reader.

AMERICAN HIGHWAYS

CHAPTER I

GENERAL HISTORY OF ROAD-BUILDING

Origin of roads. Roman roads. Roman methods of construction. Roads of the middle ages. Beginnings of modern roads. Methods of Tresaguet and Telford. Method of Macadam.

IN order to deal in an effective way with any large economic problem it is necessary to trace, at least in a general way, the history of human endeavor in relation to that particular group of needs. By such a survey the observer not only learns the importance of the matter and the ways in which men have learned to deal with it, but he also comes to see the blunders which have arisen from the lack of such understanding as he seeks to acquire. Therefore we shall begin our study as to the conditions of American roads and the means of their betterment by a glance at the history of ways of communication, gathering from this history what we may find serviceable for our own guidance.

Let us note that the first step which men make above the ancient savage estate is closely related to the progress of their desires. When they cease to be content with the simple goods which they may obtain from nature just

1

about them, when they seek by trade or war to win profit from their neighbors, the questions of transportation and the routes to be followed by it present themselves to mind. If the savage, in the first steps of his upgoing on the way toward civilization, is so fortunate as to obtain possession of animals which can be used as beasts of burden, the road problem at once opens before him. Whether it be in China, in Europe, or in Peru, he quickly learns to take the load from his own weak back and to put it on the stronger bodies of his four-footed slaves. Beginning with pack-trails following his older foot-paths, he soon learns, in a rude way, the simple arts of the road engineer. The path has to be artificially cleared and rude bridges made. Here and there, indeed, where the footing was hopelessly bad he learns to help the conditions by rude pavements.

The pack-train state of civilization may be said to be, so far as transportation is concerned, the first stage of that development. It was a stage which was long continued even in the oldest settled lands. It is consistent with a considerable advancement in culture, but not with a high commercial development. It is characteristically the method of carriage among peoples who dwell in small isolated communities, where the folk profit nothing from the life of their neighbors beyond the limits of a day's journey.

More than half of the inhabited world is still, as regards its transportation, in the pack-train state. This is true of the greater part of South America, and a considerable portion of our own continent as well. Within twenty years the writer has seen, from a county in eastern Kentucky, a caravan of small mountain-bulls, each provided with a "sawbuck pack-saddle" on which was packed the exportable product of the district, feathers, pelts, beeswax, and ginseng, which was thus conveyed over the mountains to railways.

ROMAN ROADS

The first advance beyond the method of packing was made in the regions about the Mediterranean, perhaps as early as ten centuries before the Christian era. The initial step toward the wheeled vehicle appears to have been attained in such rude tasks as moving large blocks of stone with the aid of rollers placed beneath the masses. Such work as done in Egypt must have required the construction of hardened ways. From this primitive experience with the wheel, invention appears to have led to the construction of the chariot, first used in war as a carriage for archers and javelin-throwers, but soon to come to service in the peaceful interests of transportation. In no part of the far East does the wheeled vehicle appear ever to have been extensively used for commerce; such work was done mainly with pack-animals. The greater portion of the district of Egypt and the adjacent countries is so far clear of forests and, in general, so far open to carriages that it was not absolutely necessary to make provision in the way of roads for the movement of the wheeled engines of war. It was about the time when the Romans attained strength and developed a highly organized system of administration that these carriages, modified for carrying burdens, appear to have come into general use. The first step in this direction led to the construction of a vehicle modeled after the fashion of our ordinary carts, commonly arranged to be drawn by two animals, usually oxen. Long after the invention of this simple carriage, after, indeed, good roads had begun to be constructed by the masterful Romans, a front pair of wheels was added. With this addition the cart became the ordinary springless wagon, and was, for the first time, fitted for the carriage of heavy burdens on well-constructed ways.

Although, so far as known, the Romans were the first people to undertake the construction and maintenance of costly roads, they did not enter on this field of their engineering work until relatively late in their history. It was only when their colonial possessions became large and extended into regions far from the sea-shores that they felt the need of such high-grade routes. We have little evidence to show when these ways began to be paved in the massive and thoroughgoing way which has caused them, in many cases, to survive to the present time. It is evident, however, that the Italian portion of these improvements had been well developed as early as the third century B. C. This is shown by the swift marches of the armies of the republic up and down the peninsula. A particularly good instance of these wonderful movements of large bodies of troops, necessarily accompanied by wagon-trains, is seen in the march of the consul Nero when he left southern Italy to meet the advance of Hasdrubal, who was coming to the relief of his brother Hannibal. This march was made up to the valley of the Po, and returning, a distance of over six hundred miles, at the speed of more than forty miles a day. It is impossible to believe that this movement was made over any but a good road.

With the extension of their empire, the Italian system of roads was continued in a far-extending network into what is now France and Switzerland, and even into Great Britain as far north as the Scottish border. Similar roads, though on a less extensive scale, were made in the eastern portions of their empire. It may indeed be said that with the Romans the problem of rapid and cheap land transportation was first solved. These roads were at first designed for the uses of war, to enable troops to march in compact order, and to be followed by provision trains and the instruments for hurling projectiles which served the purpose of our modern artillery. But though intended in the first place as a means of assuring conquests and administering provinces, these roads were the basis of a great commerce which served to give to the Roman empire a unity in its extension such as has been attained by few modern states.

ROMAN METHODS OF CONSTRUCTION

It is characteristic of all the engineering works done by the Roman people that they were excessively massive. Without the learning concerning the resistance of materials which we possess, without, indeed, much trace of the habit of inquiry which lifts the artificer to the grade of an engineer, the Romans built with an utter disregard as to the relations of strength and strain. This is seen in their architecture, but it is particularly manifest in their roads. As the labor involved in such works was forced, indeed generally that of slaves, considerations of economy do not appear to have entered into the reckoning, as will be seen from the diagrams (Figs. 1, 2, and 3). It should be noted that where, as in southern Europe was usually the case, the road was covered with stone blocks, these were usually bedded in the upper layer of cement. It is commonly asserted[1] that the Roman roads were built in one pattern

---

[1] "L'antique méthode des Romains était généralement de donner trois pieds d'epaisseurs à leur paré qui était composé de deux couches des pierres plats ou fond, d'une couche des matériaux plus grossiers par dessus et ansi de suite, en couches régulières, dont la dernière n'était autre q'un beton dans lequel on fixait les pierres de la surface." (See "Annales des Ponts et Chaussées," "Mémoires et Documents," 2d series, 1850, p. 60.)

1*

of construction, and that this involved a pavement of a block-like nature on the surface of the way. In fact, as is shown by Dietrich,[1] there is a considerable diversity in the method of building, the invariable features being some form of foundation made of large stones, with a layer of cement at a higher level. Where fit stone for paving in the manner of blocks was to be had, the surface was

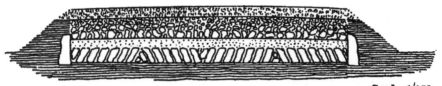

Fig. 1.

Scale 1/150.

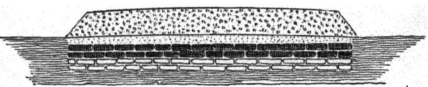

Fig. 2.

Scale 1/150.

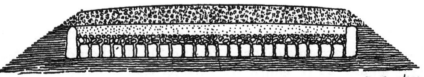

Fig. 3.

Scale 1/150.

thus covered, the masses being set in cement; where such a pavement was not conveniently possible, the surface was formed of a beton, or mortar mixed with bits of stone. Here and there, in the successive beds, layers of sand or gravel were often introduced.

[1] "Die Baumaterialien der Steinstrassen," p. 4 (Berlin, Julius Bohne, 1885).

In the construction of a Roman road we note a crude perception of the solidity which stone foundations afford, and also, in the cement layer, a recognition of the importance of keeping the road dry; beyond these half-formed conceptions there is, in these structures, no trace of engineering skill. So far as I am aware, systematic drainage of the foundations of the road seems never to have been undertaken by the Roman engineers. The bed was so massive that it in a way defied the action of frost.

The sections of the Roman roads indicate that the construction was often three feet or more in thickness even in places where experience should have quickly told, as it has taught moderns, that six or eight inches of stone would have served the purpose. In a singularly clumsy way they combined layers of different substances, one placed above another, usually with a block pavement on top, in such conditions that no beneficial effect whatever could have been gained from the accumulation. In general these roads, measured in terms of the price of labor in this country, must have cost from thirty to one hundred thousand dollars a mile. It is not too much to say that at least three fourths of the expenditure was really wasted. It is true that the brutal massiveness of the construction has enabled these ways to survive for ten or fifteen centuries after the builders passed away, but this was not the intention of the constructors. They doubtless intended to do no more than seemed to their ignorance needful. Although we cannot but admire the result, we have at the same time to recognize that the waste of human energy which it exhibits indicates one of the weak sides of that wonderful people. To it may in part be due the swiftness and completeness of their downfall. The economic strain on the resources of a folk which was applied by such ex-

ceedingly misdirected labor must have counted against
them in a time of trial.

There has been much speculation as to the manner in
which the feet of the Roman horses or oxen were protected
from the wearing to which they would naturally have been
subjected from their contact with the hard stone which,
in the form of slate or blocks, usually formed the surface
of their roads. It has been suggested that the feet of these
animals were incased in bags of cloth or leather, but any
one who has essayed such a covering as a temporary
expedient when a horse has lost a shoe has had occasion
to remark the speed with which the material wears out.
Simple as the iron shoe appears to be, it was not invented
until about the third century of our era; therefore in the
days when the construction of Roman roads went forward
with the greatest rapidity it is difficult to see how they
were made serviceable. It seems to me likely that the
need was met in a very simple way, by applying to the
paved street a layer two or three inches thick of ordinary
soil. Such a covering would enable unshod animals to do
their work without damage to their feet. Here and there
beside old Roman roads, paved as some of them are with
basaltic blocks,—a stone which an unshod burdened horse
could not tread for the distance of ten miles without
becoming lame,—we may still note the existence of depressions from which materials for the renewal of the
earthen covering may have been taken. These ancient
pits have been much effaced by the processes of change
which have acted for many centuries. They are, however, sufficiently distinct to warrant the suggestion which
has been made.

ROADS OF THE MIDDLE AGES

With the downfall of the Roman empire, even before what may fairly be termed the time of its overthrow, the road-building spirit died away. It was the product of the imperial motive. It naturally passed away when Europe was resolved into its local elements, and when each ruler held his small estate more securely when he was separated from troublesome neighbors by impassable ways. In the medieval period not only were there no Romans to undertake transcontinental marches, but there was also no commerce of any moment which went beyond a limited neighborhood. The local rulers not only abandoned road-building, but there is evidence that they in certain cases destroyed the ancient ways which might have given access to their dominions. From time to time, when great leaders, such as Charlemagne, consolidated scattered governments into something like an empire, the imperial road-building motive again manifested itself, but never in a very effective way. We have to pass by the middle ages and well into the period of our modern days, with its reintegration of the detached feudal holdings, before we come once again to a road-building period.

So recent is this change to the better economic system which has once again turned the attention of all civilized states to the construction and care of roads that we can still see, even in those parts of Europe which have the best ways, marks of the imperfect communications of the medieval time. Thus in England and in many of the older parts of this country we often find the highways very broad, the location being not infrequently a hundred feet or more in width. These broad roads were intended to serve the needs of the wagoners by affording an opportunity frequently to change the place traversed by the wheels, so as to obtain in muddy times firmer ground or to enable the driver to avoid the deep cradle-holes which develop in an ill-conditioned way. The observant traveler may sometimes note on the crests and sides of hills gully-like depressions, which are what remain of the deep creases worn on the face of the country by the abandoned tracks which were followed by the pack-trains. Over these ways the ancient commerce moved, with its burden on the backs of oxen or horses, the tail of the forward animal tied to that next behind in the procession.

BEGINNINGS OF MODERN ROADS

The first step toward the modern better roads appears to have been made in about 1775, when France began to feel the awakening of that desire for better things which unhappily led in the end to the so-called French Revolution. Under that modern Cæsar, Napoleon I, the construction of roads, which in many cases was in effect the reconstruction of the old Roman ways, was demanded for the better movement of the armies which for a score of years he had occasion to move swiftly from one part of his empire to the other. Among the first of these constructions was that over the Alps from the Rhone valley of Switzerland, a road which he had found such difficulty in traversing in his descent upon the Austrian armies in the last-named country. Probably at no time in the history of road-making has the work of bettering ways gone on with such rapidity as in the reign of that marvelous despot. Although the greater part of the present admirable road system of France has been developed since the downfall of Napoleon, the impulse in this direction was clearly shaped, if not developed, by him.   Of all the marks which he has left upon France, that of its better ways may be reckoned as the greatest of the visible monuments of his reign.

The great movement toward the betterment of social and economic conditions which manifested itself in the violence of the French Revolution affected also other countries in Europe, and led to a very general improvement of the transportation routes, so that the road-building motive can be discerned in Switzerland, in various parts of Germany and Italy, but, above all, in Great Britain. In the last-named country the military motive had less to do with the betterment of the ways than in other lands. Great Britain, which from the time of the Tudors had been more under the influence of the economic spirit than any other country, proceeded to improve her roads mainly, if not altogether, with reference to commerce.   The densely peopled condition of the island, the considerable development of manufacturing industries, and the general presence of good road-building stones led to the extension of carriageways in a more rapid manner than in any other country; so that earlier than any other the land was provided with excellent roads, though at the present time the French have attained in this regard to a somewhat more perfect system of ways.

In the revival of road-building we see at once, as in many other things, the difference between the Roman and the modern engineer. The Roman, as before remarked, the least experimental of cultivated peoples, sought strength and endurance without any reference to the cost of his constructions. The modern folk approached the problems of highway engineering with a clear sense as to

the need of a sound theory supported by carefully gaged
experience. This led the road engineers to note that where
broken bits of stone are placed upon a road to the depth
of several inches the fragments, when traversed by wheels,
soon become compacted into a solid mass, provided the
quality of the stone is such as to permit this aggrega-
tion. In this condition the pavement becomes essentially
like a slab of tolerably solid rock, through which the
wheels will not break until the sheet has worn thin by the
impact of vehicles and feet. The use of broken stone in
a reckoned minimum thickness of coating, the material
being placed upon a well-shaped bed, appears to have been
one of the many interesting contributions which the French
have made toward the betterment of roads. The beginning of this system, which has served so well, was made
about 1764 by a French engineer, Tresaguet. In his
method of construction the stone element of the road
consisted of a foundation made much in the manner of
the Roman ways, composed of large fragments of rock set
closely together, their projecting points being broken off
and their interspaces filled with smaller bits. On this
heavy foundation there was laid a covering of small frag-

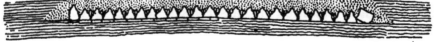

Scale 1/150.

Fig. 4.

ments such as are used on our modern broken-stone roads.
The roads built by Tresaguet afforded the pattern for the
road construction in France for nearly fifty years after
his method was introduced.

From France the improvement introduced by the French

passed to England, where, in the hands of Telford, who began his work about 1820, it underwent certain modifications, which, though at first sight unimportant, constituted a considerable advance. As ~~will be seen from the~~ diagrams, the superficial resemblance between the methods of Tresaguet and Telford is very great. So far as the foundation of large stones is concerned, there is not much difference between the two plans. The innovations in the Englishman's system consist in his arrangement of the foundation so that it shall have an arched form, following the curve which is to be given to the surface of the road. Moreover, while the method of Tresaguet provides only for a thin layer of broken stone, which apparently was in the first instance intended only to smooth over the irregularities of the foundation, that of Telford demands half a foot in depth for the upper layer, the fragments being gaged so that they will pass through a circular opening two and a half inches in diameter. Furthermore, his method requires that the stone should be placed on the road in two successive layers, the first having a depth of four inches, which, after it was trodden down by the wheels, was in turn to be covered by a final layer of the same kind of broken stone, two inches in thickness, to which was added an upper coating of clean gravel one and a half inches in depth.

METHODS OF TRESAGUET AND TELFORD

It is interesting to note in these systems of Tresaguet and of Telford the slow manner in which road construction became emancipated from the control of tradition. Working in a country which abounded in well-preserved remains of the ancient Roman ways, the highway engineers, though they began to perceive that the broken stone would compact into a solid mass, could not bring themselves to believe that a good road could, with any conditions of foundation, be made of broken stone alone. They clung to the ideal which guided the Romans, and which led them to believe that a massive underpinning of large stones, wedged together, was necessary to support the seemingly frailer upper layer. It required an inventive genius in road construction to make the last great advance in the cheapening of ways, through the abandonment in all ordinary conditions of the substructure to which the roadbuilders had so obstinately clung. This man was Macadam.

METHOD OF MACADAM

It is not impossible that Macadam obtained the first suggestion of a road built altogether of broken stone through the construction of Napoleon's road over the Simplon, where, because of the fact that nature had provided a rock foundation, the old Roman basement of the way was omitted. Macadam's work, beginning in 1816, somewhat antedates that of Telford, but, as is often the case with notable inventions, his work became more appreciated in France than it was in England, and only after many years was received with suitable favor in his own country.

To Macadam we owe the first clear understanding as to the sufficiency of the broken stone to maintain itself in a perfectly satisfactory way, quite as well as though it were on a pavement foundation, wherever the earth of the undersoil is not of soft clay lying in a position where it may readily become mud. He saw that a moderately thick layer of broken stone when well compacted would prevent the passage of water into the under earth, which when dry would afford well-needed support. Moreover, though Telford had noted that the stones which were to be compacted must all be of small size, he did not fully appreciate the fact that the matter of their diameter was one of much importance; that while the lower layer might be composed of bits three or four inches in diameter, that just beneath the wheels should be made up of pieces not more than from one and a half to two and a half inches through.

Although Macadam, as is often the case with inventors, overestimated the sufficiency of the layer of compacted broken stone to maintain its position under conditions of soft foundation, there can be no doubt that his work, inasmuch as it emancipated our road-masters from the costly ancient traditions, constitutes one of the most far-reaching inventions which have been made in relation to wheeled ways.[1] The best modern practice in the construction of country roads which are to serve a large body of travel, as well as that followed in the greater part of city streets, combines the methods of Macadam and those of the Roman type, which were brought into extensive use in England by Telford, the foundation of stone blocks firmly wedged together being used only where the under earth was of an unstable nature, or where it cannot

---

[1] The propriety of calling the process of building roads of broken stone macadamizing may well be questioned. The use of the material was established before Macadam's day. From an engineering point of view we owe to the able Scotchman little save the confidence in the material when used without especially constructed foundations; this confidence is often misplaced. Nevertheless, as the general adoption of the method of building roads of broken stone without a stone pavement was due to this master, the name is not altogether misapplied.

readily be brought to a firm state by an easy system of drainage.

With this sketch of the history of the art of road-making up to the beginning of modern practice, we shall next turn our attention to a glance at the development of roads in our own country.

CHAPTER II

EARLY AMERICAN ROADS

First American road-building period

AT the time when the English settlements in North America were formed, the methods of road-building in the mother-country were in the low state to which they had been brought during the dark ages. The main ways were, except in the dry season, unfitted for any considerable use. Much of the travel in the remoter districts was still carried on by means of pack-trains. On this account the traditions which came with the early settlers to America were practically valueless as a guide in the construction of good ways.

If the colonists of this country had come to it in the time of Rome, they would have been provided at the cost of the empire with a general system of good roads such as that masterful people usually constructed in their colonies. Although these ways would have been built for the needs of military occupation, they would at once have served for economic development, as they did in many other lands. But at the time when the American settlements were made the mother-country gave them nothing that cost money. The settlers had to provide all the framework of civilization for themselves. Bringing with them no traditions of good ways, and without the means from

2   17

their scanty earnings of making such a provision, our people began by accepting as inevitable a class of roads which was of such low grade that they have proved the greatest possible hindrances to the material and social welfare of the land. For a hundred years after the settle. ment of the Atlantic seaboard began there was hardly any way in this country fit for carriages. The intercourse between the settlements was maintained by boats, or by paths which were barely fit for horsemen and pack-animals. With the gradual increase in wealth and population the use of carriages increased. The original trackways were cleared so as to permit the passage of vehicles, and some. thing like roads began to be established. It was, however, another century before any considerable part of the traffic of the country passed over ordinary highways.

So far as the inquiries of the writer have extended, there appears no reason to believe that any well-paved roads existed in this country outside of the cities and towns until after the year 1800. When the systems of Tresaguet, Telford, and Macadam began to be applied to English ways, when it was shown to be possible, without excessive cost, to make excellent roads wherever suitable broken stone could be provided, there arose in this country a remarkable movement in favor of bettered routes. Much to develop this motive was done by the construction of the so-called National Road, which was carried from Washington westward to the central portion of the Mississippi valley.

FIRST AMERICAN ROAD-BUILDING PERIOD

At the outset of what we may call the first American road-building movement, which began about two hundred years after the founding of the first English settlements, our people in general followed not only the English method of construction, but also the plan so common in the mother-country of building these roads by the instrumentality of corporations, which were allowed to charge toll for the maintenance of the way and for the profit of the stockholders. The toll-road system did not originate at this time; it began much earlier. With the advance of the population to the westward, in the last quarter of the eighteenth century, charters, State and national, were given to many turnpike companies, who would undertake to clear away the forests and in a measure to grade the ways so that they might be made passable for strong wagons.

A notable instance of one of the early toll-ways is to be found in what was known as the Wilderness turnpike, extending from the Shenandoah valley, in Virginia, westward by way of the waters of the Upper Tennessee and Cumberland Gap to central Kentucky. This route, which, owing to the topographic features of the country it traverses, was with difficulty opened to the passage of pack-trains, was by the corporation which undertook its improvement brought into an excellent condition for use by wagons. It was, indeed, the main way by which the people of Virginia found a passage into the valley of the Ohio. The early construction of this road permitted a relatively large amount of commerce between the East and the West, and allowed the settlements of Kentucky to attain by the beginning of this century a strength which enabled them rapidly to make head against the difficulties of the wilderness and the resistance which the Indians and the British opposed to the advance of the American people in that part of the world.

In Virginia, Kentucky, Tennessee, and in parts of the

neighboring States, the absence of the encumbrance of glacial drift, and the general presence of good building stones, made it possible to develop the system of macadamized ways, built by corporations, and maintained, often with great profit, by the charge of toll. This method of building roads, though in the end when a community becomes well established disadvantageous to its industries, is in the stage of settlement undoubtedly beneficial, for the reason that it induces capital to forerun the needs of the population, giving ways which promote the development of the land before the settlers are of their own means able to make such provision. The speed with which good lands of the States which were characteristically the seats of toll-roads were brought under tillage must be attributed to the extended use of such corporation ways.

Although toll-roads were most characteristically developed in the States above mentioned, and there remain to this day a feature in their economic system, such roads were also common in all parts of the United States in the early part of this century. Even in New England many of the greater ways of intercommunication were originally built under the toll system. A half-dozen or more of these main ways radiated from Boston, Massachusetts, and are still generally known as turnpikes, a name which marks their original conditions. In general, however, provisions have been made in the laws of all the seaboard States north of Virginia, and to a great extent in all the so-called Northern States, which have cleared away the turnpike companies' rights in the roads they once possessed, returning these ways to the free use of the public. The greater economic advance of the non-slaveholding portions of this country in the time before the Civil War is well indicated by this change in the highway system. So far as the writer has

THE FARMERS' SLOUGH.

The main road between Cleveland and Warrenville, Ohio, about two miles from Cleveland city limits, April 7, 1891

2*

learned, there are now perhaps not more than three or four toll-roads in New England, and perhaps not more than that number of bridges where toll is to be paid. These exceptional cases in which the ways remain obstructed are all under peculiar conditions which for one reason and another justify the ancient method of charging the traveler a fee for a peculiar privilege.

In the greater part of our frontier country the process of road-making, as the settlements advance westward, has in the first instance consisted in granting a location for the road along the most practicable way for use. In the beginning those who used the roads were compelled to make them fit for their needs as they might be able to do. As the local governments were organized, small appropriations were made to better the least passable portions of the route. Gradually there came to be some appropriation of money or of labor in the way of a road tax, often, as noted below, in the form of labor, which was expended almost never on a better class of improvements, but in practically all cases in making temporary provisions to keep the roads passable at least in certain seasons of the year.

As before remarked, the traditions which our people inherited in matters concerning road-building from the Old World represent the most degraded state of such public work. Included in these evil customs it is interesting to note the existence of a remnant of feudalism in the forced labor on highways which at one time and another, and still generally in many of our States, is commonly demanded of all able-bodied men who dwell in the country. This custom is sometimes called the working out of the road tax, a term which denotes a system whereby the people may pay a definite charge either in money or in labor. In

2*

other regions, notably in Kentucky, the work is said to be done by the "militia," a phrase which probably denotes the ancient feudal method by which all the men fit for war could be called on to render such public service.

There is probably no other feature in our road system which has so far served to maintain the low state of our American road-making as this "corvée," or forced-labor system on the highway. It has bred, in a systematic manner, a shiftless method of work; it has led our people to look upon road-building as a nuisance. There is no situation in which the American workman makes so unsatisfactory an appearance as when he is endeavoring to do the least possible amount of labor which is to count as a day's work on the highways of his district.

Although the working out of the road tax has been abolished in perhaps half of the country, the method is still in use in many of the Western and in nearly all of the Southern States. There is a singular custom which is followed by the road-masters in Virginia and some other Southern States, who have, or are supposed to have, the control of the impressed road gangs. It consists in cutting with an ax or knife three vertical lines in the trees from point to point along the road, these usually being crossed by horizontal lines, so as to make a rude sign for the Roman numeral III. Although there is no tradition as to the origin of this usage, its wide distribution and the fact that before the Revolutionary War the main roads were known as the king's highways makes it seem likely that the marks were intended to signify George III; there having been no successor to that monarch in this country, there has been no occasion to change the sign.

In the present condition of this country the resources which favor distant transportation are well organized.

SUBURBAN DESOLATION.

Wagons abandoned in deep mud near intersection of Ogden Avenue and 22d street, Chicago, April 6, 1901

The development of the railway and interior steamboat transportation has provided for these needs in a measure which has been attained only in some of the richest European states. It is otherwise with the ways which serve for local intercourse; these have been so far neglected that their ill condition operates as a distinct check on the social relations upon which the character of our local communities intimately depends. The political life of our commonwealths, as well as their economic advance, is to a great extent determined by the readiness with which the people obtain that association with one another which leads to the development of a public spirit. Important as are the effects of good acquaintance in the communities of any state, whatever be its system of government, they are particularly important in a democracy; for there the unending task of holding fast to the good which has been won, and of winning gains for the future, is to be effected only by means of an intense social life such as will give the able men of each neighborhood an opportunity to affect the motives of their weaker fellow-citizens.

The reasons above given should make it evident that the interest of a democracy in good roads should rest on a deeper foundation than mere commerce or commercial needs. Account should be taken of the value of these ways of communication to a people from the point of view of their place in the intellectual development of their communities. Thus viewed, good roads will be seen to have a very important relation to the mechanism of a democratic state.

The greatest hindrances which have beset the development of American roads arise, in the first place, from our system of government, which has not provided authorities competent to organize and control the methods of constructing and maintaining roads in our commonwealths, and, in the second place, from the character of the climate, topography, soil, and underlying rocks within the limits of the United States. As any well-considered movement for the betterment of our ways must take account of these difficulties, we shall now proceed to consider them, beginning first with those of a natural sort.

A load of hay in Normandy.

CHAPTER III

EFFECT OF THE CLIMATE

Effect of the under earth on roads.   Influence of the topography on roads.   Effect of forests on roads

AS a roadway is of all constructions the most exposed to the action of the weather, the climate of the district in which it lies has a greater effect upon it than upon any other class of buildings.   This effect is exercised by the rainfall, changes in temperature, and the winds.   A secondary influence, arising from the above-mentioned natural conditions, is found in the character of the vegetation, which under favorable conditions may advantageously affect a road by covering the unused portion of its surface with a network of low-growing plants, such as the grasses.

Under any conditions a road has to lie open to the rain. Where this comes gently, as is usually the case in Europe, it may not wash the surface of a well-graded way in a serious manner.   When, however, as in this country, the rainfall, particularly in the central and western portions of the land, often comes in a torrential manner, the effect is, even on well-constructed roads, to wash out the dust which holds the stones together as well as to remove the divided portion of the rock, which should have a coating to keep the wheels and the shoes of horses from breaking the stone in a rapid manner.   Thus the result of occa-

29

sional heavy rains is in this country a more rapid wearing of the road-bed than occurs in the Old World. It is doubtful, indeed, if the Roman ways would have survived in this land in the manner in which they have endured in the regions where they were built.

In almost all instances the ditches on either side of a road have to receive a large share of water which flows over the surface toward the way. Where, as in America, the rainfall may amount, as is often the case, to an inch or more an hour a large part of the water, especially when the ground is frozen, flows over the surface, and much of it finds its way to these ditches. As will be noted hereafter, the waterways beside roads are an important part of the construction. The cost of their provision and maintenance is on the average much greater with us than in European lands. Furthermore, it is essential that the earth beneath a Macadam way, where it is not provided with a pavement foundation, should be kept dry. It is desirable, indeed, in all cases that it should be protected from the invasion of water. The expense of underdrainage, such as is hereafter to be described, is exceptionally great in the case of American constructions.

The well-known heaving action of frost, which is proportionate to the depth to which it enters the soil and to the water contained therein, is always a menace to the preservation of a roadway. This movement not only disturbs the whole construction, but it tends to force up the larger stones through the macadam or gravel, so that they disturb the bed in their ascent and encumber the way when they appear at the surface. In the Northern States of this Union, where the frost often enters the earth to the depth of three feet or more, the effect of freezing and thawing, often repeated several times in the course of a winter, is exceedingly injurious. To guard against it, it is necessary to provide for the removal of the water to a greater depth beneath the surface than is required on the continent of Europe or in Great Britain.

The evils arising from the long-continued droughts which are so common in America are felt in several different ways. Where broken stone is used as road material it is held together by the cementing action of the dust which lies between the fragments. Where the way is traversed by heavy wagons it almost always undergoes a certain breaking up of the bond. This is restored by a recementation process, which causes the dust when wetted once again to bind. It thus comes about that a road which is wetted at intervals, say no greater than a fortnight, will remain in a firm state, while, when subjected to traffic for a drought of a month or more in duration, it will be broken into a mere rubble. A conspicuous instance of this action came under the observation of the writer in the campaign of 1862 between the armies of Bragg and Buell in Kentucky and Tennessee. It was a season of remarkable drought, little or no rain falling for the term of seventy days. During this time the Macadam roads of that district, which ordinarily are in an excellent condition, were by the wagon and artillery trains brought almost to a state of ruin. The fragments of stone which ordinarily adhered firmly to one another were converted into pebbles, which ground up under the tread of the wheels. It was not until after the great rains which came on the night of the battle of Perryville that these roads began to return to a fairly passable state. Many of them, however, were so injured by the grinding up of the loose fragments that they were unserviceable until they were recovered with broken stone.

The effect of the winds on roads is to blow away the protecting covering of dust. If they be strong the action may go so far as to remove the cementing material from between the exposed crevices. In general it may be said that the wearing of a road increases rapidly with the speed and continuity of the winds and the extent to which they blow in times of drought. The strong southwest winds so prevalent in this country, particularly in the Mississippi valley in the summer, much increase the cost of maintenance of good ways.

In a moderately humid climate, exempt from continuous summer droughts, creeping plants, nourished by the dust from the roads, which in most cases has a considerable fertilizing value, take hold on the shoulders and sides of the way in such a manner as to protect those exposed parts from washing or from the action of the winds. Where these conditions prevail it is generally practicable to build a relatively narrow hardened way with wide shoulders on either side on to which the passing teams can turn out, finding there, by virtue of the plant covering, a surface so firm that it will not rut from an occasional passage of wheels. If, however, the shoulders are overdry, as they are sure to become in an enduring drought, the plants are killed and the surface left unprotected.

The result of the above-mentioned climatal conditions is to make the construction and maintenance of good highways a matter of greater cost in the new than in the Old World. The conditions in the two realms are so far diverse that we need to be careful in adopting without revision the methods which have been successful beyond the Atlantic. In all cases these methods should be critically examined with reference to the climatal and other needs of this country.

EFFECT OF THE UNDER EARTH ON ROADS

The character of the under materials of a country profoundly affects the problem of road construction in two ways: as concerns, first, the nature of the foundation, and, second, as to the sources of supply of the materials which are to be used in the hardened way in paving gutters, and in drains. Where the hardened way can be laid upon a base which holds but little water, so that it is not liable to become soft or to move in the manner of loose sand, the problem is relatively simple. Where, however, as is often the case, the foundation is of a more or less plastic clay, of muck, or of yielding sand, the precautions which the road-master has to take add much to the cost of the construction. In general it may be said that the difficulty arising from the nature of the underlying materials of this country adds much to the expense of building good roads. Wherever the soil is deep and therefore of a high order of fertility, and from the fact that such a deep soil means in all cases a considerable proportion of clay and a ready penetration of water into it, road-making is exceptionally costly for the reason that some provision similar to the Telford pavements has to be made to keep the surface coating of broken stone or gravel from working down into the bed. It may be said, indeed, that the natural fertility of the land adds much to the cost which has to be incurred in providing highways, by giving the conditions which bring about bad foundations.

INFLUENCE OF THE TOPOGRAPHY ON ROADS

The surface of a country, the relations of the hills and valleys which go to make up what is called the topography, profoundly affects the cost of roads and requires a peculiar skill in planning the line which the way is to follow. It is desirable that the line adopted should have a sufficient grade to remove the water from the surface and ditches, and that such grades should be so varied that the draught animals may not have an unvaried burden. Any slope beyond that required for the removal of the water is a hindrance to transportation which increases at a very rapid rate with the steepening of the declivities. It is, moreover, in a high measure desirable that the main way should be so placed that the territory which it serves shall, as far as possible, have the auxiliary ways sloping toward the main route. Among the many evils brought about by the lack of proper engineering skill in the location of American roads, their ill position in relation to the country which they are to serve is perhaps the worst.

In planning the route which a road is to follow, or in remedying by changes of position unfortunately placed existing ways, it is necessary to take into account the character of the topography of the district. This varies greatly in different parts of this country. In nearly all the region which has been in recent geological times covered by a glacial sheet, the surface of the land has a curiously irregular form. While the main valleys, those which were formed before the ice invasion, preserve their outlines in a general way, the paths of the small streams and the table-lands between them have been covered by irregularly disposed masses of debris which give the surface a broken character which requires great care in placing the road in order to avoid unnecessary grades. Moreover, in this part of the land the underlying materials are extremely variable in their nature, sometimes being very clayey, and again very sandy, and it is therefore necessary for the road-master to study the distribution of the materials on which the road is to be founded.

The portions of this country which were affected in the manner above described by the glacial action may be denoted as follows. Beginning on the middle portion of the Atlantic coast of New Jersey, all the country north of that line and its continuation to the westward bears the ice-mark. In its western continuation the southern limit of the ice-sheet extends across Pennsylvania to the head waters of the Susquehanna and the Allegheny rivers; thence, with many local variations, in a general way down the northern side of the Ohio, some distance north of that stream, to Cincinnati, where it crosses the river, barely entering the State of Kentucky; thence westwardly through Indiana and Illinois to the Mississippi River, and along a much-varied line up the valley of the Missouri to the far West. In addition to the great field lying to the north of the above-described line, there are in Colorado and on the Pacific slope of the continent various areas which in a less considerable manner have been affected by the glaciers. Throughout the glacial field above described, though in a varied measure, the location and foundation of roads are much affected by the irregular distribution of the glacial waste.

South of the regions affected by the last glacial period, all parts of our country, except certain limestone districts where "sink-holes" occur, exhibit the feature of continuous slopes from the head waters of the streams to near the level of the sea. Moreover, in this part of the realm the underlying earth materials which may affect the foundations of a road are much more uniform in their character than in the region first noted.

While in a glaciated district the road-master needs carefully to inform himself as to the shape of a country which the road is to traverse, so that he may take advantage of each feature in relief or in the under structure, in the non-glaciated district he may in general plan the roads so that the main ways pass down the greater valleys or, if they be not readily passable, along the divides.

In adjusting a road with reference to the topography and the underlying conditions, there is, as has been noted by any student of roads, an exceeding diversity of method. It was the custom of the Romans, due to the rude forceful motive which they brought into all their architecture, to carry their ways straight across the country, rarely deflecting for any but absolutely insuperable obstacles. This humor curiously returned among the Puritans of New England, who appear to have been disposed to deal with the difficulties of the earth in a like direct manner. They too built many of their roads, particularly the turnpikes, straight across the country without any effort to adjust the routes to the topography. With this exception, in all modern road-making a great and increasing advance has been effected in the adjustment of the ways to the surface. This end has been most completely attained by the Swiss road-builders, who have given us routes crossing their mountains which are masterpieces of adjustment so contrived that the transportation may be effected with combined speed and ease.

In this country, although the discerning observer can often note the work of masterful road-makers in the adjustment of ways, the results almost everywhere show a lack of the contriving motive which is elsewhere characteristic of American construction. The writer, who has noted the conditions of the roads in many parts of this country, has rarely found, in a region of varied topography, any stretch of five miles in length which did not exhibit some glaring error of position which if corrected would have materially increased the value of the way. In many cases it is patent that the wrong placement of the route has been due to the influence of some landowner who has secured a personal advantage at the cost of all the other folk who may be interested in the route. It is perhaps impossible to avoid such influences, but on their avoidance depends the rational planning of our roads.

Supposing that a road-master has at his discretion to lay out a way or to remedy the defects of one that exists, the following method of procedure may be recommended. If possible a good contour-map of the region should be obtained: one which will exhibit, by lines drawn around the hills and valleys, the place of water-levels, say at the height of ten or twenty feet, one above the other. On such a map a project for a route may be devised by provisional lines so laid as to escape steep grades, at the same time avoiding excessive length. The course to be taken can, in a general way, be readily computed from the intervals between the levels indicated by the contours and the length of the lines; i.e., the length of the proposed road may be computed in any one of several easy ways. In this work it is necessary to take account of the present tillage and other interests of man which relate to transportation. A road may often fitly be lengthened by many per cent. in order to accommodate the existing or anticipated travel. Here, as in much other work, the discretion of the road-master has to come in in a way for which no prescriptions can be given. It may be said, however, that the extension of farming in an undeveloped country, the use of water-power, the opening of quarries or mines, the probable growth of towns, all have to be taken into ac-

3

count. Even more difficult to reckon is the use of the way in distant transportation, that which comes to it from beyond the points of the field which it in an immediate way serves.

A project formed in the manner above noted should next be criticized by a close study with the surveying instruments and by a study of the topography in reference to dealing with the water which is likely to come upon the road or which has to pass under it. This study should be extended to the underlying materials so that a computation may be made as to the need of costly foundations. In parts of New England roads have been carried across swampy places without any reckoning as to the cost of maintaining them in such positions. The result is that many of them have continued to sink into the soft material until very costly fillings have been made, all of which could have been avoided by a slight deflection of the way. In other cases patches of soft clay, such as are likely to be found in glaciated countries, and which greatly increase the cost of building a good road, have been traversed, though they might easily have been avoided. Yet further, in almost all glaciated districts and to a certain extent in those of ordinary topography, a good road cannot be made without much cutting and filling in order to insure reasonable grades. Care should be taken to bring these cuts and fills into such relation that the earth which is excavated can be used in bettering the grades by filling. Moreover, where cuts are required care should be taken to place them as little as possible on steep side-hills, so that the continual slipping of the earth down the slope may not prove a source of permanent cost.

Quite as important as the adjustment of the way with reference to the immediate conditions are the problems re-

VIEW ON HUNTING PARK AVENUE, PHILADELPHIA.

About four miles from the City Hall, February 23, 1901.

lating to the materials which can be used in the construction and repair of the way, as well as the other expenses of maintenance.

As will be further noted in the chapter on road materials, by far the largest part of the expense incurred in bringing an existing or an old way into good condition arises from the cost of the materials used in hardening the way. This expense is incurred not only in the original construction, but in the constant repairs. It may be estimated that at the end of fifteen years a road, sufficiently used to warrant a carefully made stone or gravel structure, will have to be completely renewed by repairs and reconstructions. Therefore the expense of obtaining the materials is great, and has to be considered in the choice of the location. In a country of varied underlying rocks, such as is found in New England, in the greater part of the Appalachian district, or in the field to the westward of the central plain of the continent, it will often happen that a slight deviation of the way will enable the road-master to command the resources for building and repairing in a measure not possible on a shorter route.

Where, as in a large part of this country,—as, for instance, throughout the greater portion of the Mississippi valley,—the rocks lie in nearly horizontal attitudes, it is desirable to keep the road location so far as possible in a position such that the immediately underlying rocks may afford the most suitable road material of the country. Where, as in New England, New Jersey, and in many more southern parts of the Appalachian district, the sources of supply of road material are from the rocks commonly known as trap, it is very desirable that the position of these deposits should be ascertained before the location of any important way is determined on.

An analysis of the location of American roads, such as a student of these constructions may readily make, discloses two general methods which are followed in laying out these ways. In the one an effort is made to keep the route on the elevated lands between the main streams. These may be termed divide roads. The other is to place the ways in the valleys. As between these two rules of location it may be said that the divide roads have in general the advantage of dry foundations. They, moreover, escape the costs of dealing with streams in their ordinary state and with flood waters. They can in many cases be made more direct between chosen points. The disadvantage of such locations is that, unless the table-land be very wide and the valleys narrow, a large part of the transportation has to be uphill over the necessarily poorer roads which lead from the farms to the main way. Moreover, in times of drought such divide roads are apt to become excessively desiccated and to lack water-supply for beasts of burden. A conspicuous instance of this occurred in the before-mentioned campaign of 1862 in Kentucky and Tennessee, where during the great drought many of the main highways, which in that region lie usually on the table-land divides, so far lacked water that it was often necessary to march thirty miles or more from the direct path in order to obtain a supply for the columns of troops.

Valley roads, while they escape the disadvantages above noted, incur those of bad foundations, of much bridging, and of injury from floods. It may be here noted that where flood waters, even of very slight current, pass over a macadamized road, they, by removing the cementing material,—an effect which takes place to the depth of some inches below the surface,—are likely to reduce the way to a state of rubble, which will not recement until subjected to careful repair. Even the tread of the wheels, which will "bring down" newly laid broken stone, has little effect on the partly rounded bits which have been washed clear of the cementing material. The loss of their angles which takes place in the process of first bedding is apt to make the fragments shear under the tread of ordinary wagon-wheels, which otherwise would have forced them into a firm state.

The foregoing considerations warrant the following statement concerning the location of highways, which, though general in its nature, will have some value as a guide in the location of roads. Where, as in most regions of a table-land form, the valleys are narrow and the uplands broad, with a slight fall toward the cañon-like seats of the streams, the roads had best be organized in relation to the divides, leaving the tillage areas in the gorges to bear the tax of the difficult transportation to the upland. On the other hand, where, as is the case with the greater part of this country, the divides are narrow and the greater part of the culture is adown the slopes on either side from them, the roads had best be planned in the bottoms of the valleys or, if they are much subjected to inundation, on the sides of declivities above the flood-plain.

In placing the roads on the divides it is often good policy to avoid straight lines, even if such be practicable, and to bend the road downward at the head of each considerable valley, so as to minimize the cost of drawing freight over the poorer class of sideways. If carefully planned the grades thus brought about in the main way will not be greater than is desirable to afford relief to the draught animals from the unvaried and peculiarly destructive effect arising from drawing a burden on a level way.

3*

In valley roads abundant experience shows that where possible road constructions on the alluvial plain should be avoided, for on that kind of ground the foundations are almost invariably bad. The gorges cut by the streams are usually wide and afford a poor base for bridges, and the way is remote from road-building stones, except they be brought in by railways. The best place for such a road is where the hill slopes come to the margin of the alluvial plain. In such positions the under earth is generally occupied by talus material, a rubbly mass which has worked down from the hills and which affords an excellent foundation. Moreover, in this position the stream gorges are usually narrow and not very deep. The water which courses in them has a high speed, so that a relatively small way will afford it passage.

EFFECT OF FORESTS ON ROADS

The effect of forests on the construction and maintenance of roads is considerable. Where these woods are deeply rooted it is necessary to exercise a considerable amount of care in removing the woody material, not only the crowns and tap-roots, but also those of any size which penetrate downward, and this for the reason that the decay of the remains of the tree is apt to bring about harmful settlements of the foundation. The cost in general of carrying a road through thick woods is, so far as the preparation of the bed is concerned, at least twice as great as where it traverses an open country.

The effect of a timber belt on either side of a road is sometimes to necessitate more careful drainage to insure the dryness of the subway. Where the hardened part of the construction is made of gravel the influence of the shade and of a plentiful contribution of fallen leaves is to preserve the layer from the excessive dryness which is likely rapidly to desiccate the surface of the wheelway. Moreover, the covering of leaves affords some protection against the impact of tires and hoofs, while the result of the decay of vegetable matter is to favor the cementation of the bed.   In a less degree the shelter of a wood or of thick plantations on either side of the road, even that which is afforded by the ordinary spaced trees which are commonly planted beside ways, is helpful to Macadam roads.   Trees also diminish the ill effects of winds, retaining the dust on the road in a way that it would not be kept there if the road lies quite open to the blast.   On these economic accounts, as well as for the grace which plantations afford, it is advisable to keep a way tolerably shaded, at least in such a climate as exists in almost all parts of the United States.   Further consideration of this matter will be given in the pages devoted to the adornment of roads.

CHAPTER IV

NATURE AND DISTRIBUTION OF ROAD MATERIALS AND THEIR METHODS OF USE

Farm roads. Neighborhood roads. Main highways. Road-making materials in general. Trappean rocks. Granitic rocks. Quartzites. Limestones. Cherts. Clay slates. Gravels. Glacial gravels. Stream gravels. Boulder deposits. Conditions of glacial deposits. Gravels of the Southern States. Phosphate nodules. Shell beds. Paving-brick clays

FROM the point of view which is now to be taken roads may be divided into four groups, according to the amount of traffic which they are ordinarily called on to bear. The lowest of these grades may be termed farm roads, which cannot be the objects of much expenditure. Next are the public ways which serve limited districts and which, though the subject of municipal care, cannot be made in a very costly manner. The third group are the main arteries of a country district which serve for intercommunication between distant points. The fourth group includes the costliest, though the least common—the great ways which lead to great commercial centers and merge into ordinary city streets, which streets are not considered in this writing. The materials to be used on these several classes of ways necessarily differ.

FARM ROADS

On the farm roads the means for construction and repair have naturally to be sought near at hand. It is not practicable to spend much money in preparing them for use. It is not usually possible to incur much expense in drainage. In farm roads, because of the fact that it is rarely practicable to keep the surface in a smooth state, it is desirable carefully to consider the matter of grades and to maintain the surface in such a condition that it will not rut to any depth. Where a stone fit for road-making in the manner of Macadam can be had at small cost, it will generally be found profitable to cover the main farm ways with such material to the depth of six inches and to the width of eight feet. When this cannot be done and gravel is obtainable, it should be used to the depth of eight or ten inches, the material, if possible, being first screened when dry, so as to separate the clayey matter which it may contain. Where neither of these resources can be had, the only protection to be obtained is that from some form of vegetable matter.

It is easy to note, where a wagon passes over an old sod, that the wheels are upborne in a tolerably perfect way on the first passage of the vehicle. Thus a heavy wagon-load of hay may be drawn over an ordinary meadow, leaving so little imprint that it will not be visible the next season. The fact is that, until broken up by the repeated tread of the wheels, a thick coating of sod affords a sufficient support for a tolerably heavy vehicle. It is possible on many farm ways to make use of such a sod as a road covering. Where ruts form to any depth they can be closed by cutting a wedge-shaped section from the sod on either side, which has commonly been forced up above the trough, so that the channel may be closed. If this is done in the springtime, the detached sods rammed into place, and the surface strewn with any fertilizer, the effect is quickly to restore the turf. This protection will not endure a large amount of traffic, but, except in very dry regions, the advantage of the method has been proved by experiments which the writer has made.

If the farm roads be in a clayey district the ruts may advantageously be filled with gravel without any effort to cover the whole surface of the road. Broken stone may also be made to serve the same purpose, but it should be composed of relatively small pieces, preferably none over two inches in diameter. The filling should be done in dry weather and, if possible, the mass should be rolled down by means of a broad-tired wagon carrying a heavy load. Although this work had best not be done when the roads are muddy, for the reason that the stone will churn and so lose its binding property, it is to be observed that the mass should, if possible, be watered before rolling in the manner above described. The stone may be put on in dry weather and rolled after the first shower, thus saving the cost of artificial watering. If the rut be more than four inches in depth it is best that the stone should be put in as two successively applied layers, the lower being rolled before the second is placed upon it.

Where stone is accessible on a farm, and well-built roads are out of the question because of their cost, the ways may be cheaply bettered, if not made good, by plowing and scraping the track to the depth of six or eight inches and to the width of eight feet, filling the trench with large stone to the depth of about six inches. This layer should be roughly compacted with a paver's rammer. On top of this layer other stone should be placed in succession, the pieces being broken with the hammer or sledge. As the amount of "fines" or dust thus produced is insufficient to cement the bits together, gravel to serve as a "binder" should be sought for. A layer of this material about an inch in thickness may be laid on the surface. The use of clay for this purpose should be avoided. If the rock be a limestone it will cement tolerably well without any binding other than that which will soon be provided by the action of the wheels.

In clayey soil the above-described plan should include cross-drains at the lowest points so arranged as to keep the water from the foundation of the road. Care should be taken that the water does not flow for any distance in the road. As a regular crown is not attainable in such a rough-built way, the checking of the flow will have to be accomplished by cross-ridges after the plan of the Yankee "thank-you-ma'ams." These must not be made sharp-crested or the road will inevitably go to pieces on their summits. For a height of one foot they should have a cross-section of about ten feet.

In regions where neither stone nor gravel can be obtained, especially where the earth is very sandy, as on Cape Cod, there are two other ways in which the cheaper class of roads, particularly those for farm use, may be benefited. One of these, more or less in use in southeastern Massachusetts, is to cover the road with turfy material, preferably that composed of the roots of huckleberry plants and similar low-growing bushy vegetation. In practice the best method is to cut these sods into strips about eight inches wide, applying them in a vertical position somewhat in the manner of paving-bricks, pushing the sheets close against one another. A road thus made, though by no means a first-class way, may be tolerably smooth and moderately enduring. In a similar manner partly decayed leaves and, indeed, any vegetable fiber, such as sod, may be incorporated with the loose earth to the great advantage of the road. The writer has seen the streets of a town in northern Wisconsin brought from an impassable to an excellent, though temporary, condition of repair by covering the whole surface with a layer of shredded wood, commonly known to upholsterers as "excelsior," to the depth of ten or twelve inches. The result was a springy road, which, with the addition of a little clay or even that brought in on the wheels of wagons, speedily formed in tolerably smooth, elastic pavements. It was stated that such a construction would prove serviceable for a term of five or six years. Such a road covering would be objectionable in the streets of the town, for the reason that it would soon become, from the dung of horses and cattle, a mere hotbed for the development of germs. It is possible, however, that it may advantageously be used in some districts where nature has provided no other resource.

NEIGHBORHOOD ROADS

The ways of the second order, those which the French call vicinal routes, generally require, because of the increased traffic, much more extended care than farm ways. Where suitable broken stone for road-making can be found a careful reckoning shows that it is, considering the costs of repair, often advantageous to build these roads in the Macadam manner, using the Telford foundation where such may be necessary. If the foundation be very sandy, unless the sand has a somewhat cemented character, it is well to provide against the mixing of the broken stone with the loose underlying material by some one of several devices. Where possible this end can be attained by bringing the way into a grassed state. Where the road is of suitable form and grass-covered the natural foundation should not be disturbed by the stone that is to be placed on the top of it. The writer has found that where the stone is to go upon loose sand it is economical, after shaping the sand to the curve of the road, in the manner indicated in the chapter on road-making, to cover the bed with one thickness of cheese-cloth, such as may be had for not more than four cents a square yard. If carefully strewn upon this cloth, broken stone will not cut it through, and the mass may be brought into position by a roller in the usual manner of making Macadam roads. On the neighborhood roads here under consideration it is not usually desirable to have the broken-stone covering more than eight feet in width, pains being taken to provide intervisible turnout points where vehicles may pass each other. If a road has only the ordinary traffic of an agricultural district, the aggregate of teams passing does not usually exceed forty in a day. Under these conditions, a narrow road, such as is above described, will well serve the needs. No serious inconvenience will be found from the narrowing of the way, save that in the night the drivers of wagons may not be able to perceive each other at a sufficient distance to make sure of a turnout place.

Next in value to broken stone as a material to be used on neighborhood roads is gravel or natural broken stone, the fragments of which have been rounded by water. Where a good "binding gravel" can be found, one that binds because the materials tend to adhere together through the presence of iron, lime, or other cementing material, good ways may be built of it. As before noted, the material should be screened. It should be placed in a layer not less than eight, and preferably ten, inches thick. It should be given close attention in the matter of repairs, which it needs much oftener than a way of broken stone. In making such gravel roads the roller is unnecessary. After the surface has been brought into form, the gravel may be trusted to come into a firm state by the action of wheels. It is in certain cases desirable, on such a road, to sow in grass the parts not continually trodden, with the addition of fertilizing material to insure swift growth; and this for the reason that such vegetation as may survive the effects of travel will serve to prevent the blowing of the dust and also to contribute vegetable matter, which, by its decay, helps in the process of cementation through the chemical changes it produces in the stony matter.

MAIN HIGHWAYS

The main ways of a country district should, wherever conveniently possible, be made of broken stone in the manner of Macadam. As will be set forth in more detail in the chapters on the process of road-building, such roads require a hardened way, which cannot well be less than twelve feet in width and usually does not need to be more than fifteen feet wide. On such a road it may be safely estimated that the amount of broken stone required will vary according to the weight of material, which differs, in various species of rocks and according to the depth which it may be necessary to use, from twenty-two to thirty-two hundred tons. _per mile_ (See Appendix A.) On the narrower hardened ways recommended for the neighborhood roads the amount of the stone used need not exceed about one half these quantities.

On the highest-grade country roads, those which immediately connect with the great arteries of cities, pavements made of broken stone are apt to prove insufficiently enduring for heavy traffic. In these cases it is often necessary, or at least cheaper, to pave the road with stone blocks. In such instances the road may, even if it be in the country, be regarded as a part of the city streets. As such it is only mentioned here to show the classes of road materials, the distribution of which we have shortly to consider. In addition to the gravel and stone above described there are certain parts of this country on the seaboard where oyster-shells or similar materials may be used in road-making. These materials are, however, limited to the coast belt. They will receive but incidental mention in this chapter.

The brief statement concerning road construction, so far as the materials are concerned, which is given in the foregoing paragraphs, makes it clear that the quantity of these substances required in the construction and maintenance of the roads in any cultivated district is very great. Thus on the main ways of the small State of Massachusetts, which in its eight thousand miles of area has about twenty-two hundred miles of road deemed of sufficient importance to be reckoned as worthy of care by the commonwealth, the amount of stone annually needed to keep the routes in good order may be roughly estimated at about one hundred and fifty thousand tons. When the culture of this country becomes organized to the plane attained in France, it seems not unlikely that the amount of road-making stone required each year will exceed one hundred million tons. It is therefore evident that the question as to the source of supply of these materials and their relative values is a matter of national importance. We will therefore now proceed to take account of the nature and distribution of these road-making materials.

ROAD-MAKING MATERIALS IN GENERAL

Considering as of first importance the relative value of different species of rock which may be used in road-making, we will first take account of these variations in quality and then briefly set forth the geographical distribution of the various kinds.

In road-making in the modern practice we have to rest our work on the observation, which we mainly owe to Macadam, that broken stone when subjected to great pressure, as by the wheels of heavily laden wagons or steam-rollers, becomes compacted. Macadam and the other engineers of his day gave us no account as to the nature of the process by which the fragments of stone were firmly bound together. The early engineers appear to have supposed that the bits were entangled with or in a way felted the one to the other. The writer, in connection with the work of the Massachusetts Highway Commission, and from the experiments made in the laboratory of highway engineering in the Lawrence Scientific School of Harvard University, has become convinced that this process of cementation which gives solidity to the Macadam road is mainly due to, and to be measured by, the energy of cementation of the dust on the broken stone, either that made in crushing the material before it is applied to the way or that produced by the rubbing of the bits together, which is brought about by the action of the roller or of wagon-wheels; furthermore, that the binding action is, as to its value, determined not only by the intensity with which the particles hold together when first set, but by the extent to which this dust may recement when broken up by the wheels, after it has been watered either artificially or by the occasional rains. To this quality of the dust which serves to bind the stones together we have to attribute the greater part of the value which is obtained by the broken-stone method of construction. Close to it in importance is the resistance which the separate bits afford to the crushing action of the wheels. Although the dust which naturally accumulates, and which should be retained to a moderate thickness on the surface of the Macadam way, receives much of the impact of the wheels and feet of carriages and horses, the principal resistance is found in the strength of the separate bits which lie next the surface on the hardened way. If these bits crumble too readily, the road becomes a dust factory. A large part of the fine material is inevitably swept away by the rains and winds, or has to be removed by scrapers to keep the road in reasonably good condition in wet weather. Therefore we see that there are two qualities needed in a road-making stone: the dust must have a high cementing and recementing quality, and the larger fragments must be tough so that they may not be shattered by blows inflicted on them.

Experience in this country shows that the road-building materials which are generally and extensively available may be, as regards their useful qualities, placed in a series indicating the relative values in the order shown in the table, where the uppermost is the best.

> Trap.
> Syenite.
> Granite.
> Chert.
> Limestone (non-crystalline).
> Mica schist.
> Quartz.

Although all the above-mentioned stones vary greatly in quality, so much so, indeed, that the table as given has

4

no more than a very general value, it is clear that the first in order of utility is the group of volcanic rocks appearing in the form of dark-colored, massive stones, commonly known by the names of trap or basalt, terms which include a number of rock species which are discriminated by petrographers, but which cannot be separately treated except in a recondite way.

TRAPPEAN ROCKS

It is characteristic of all the trappean rocks that they have once been fluid from heat and while in that state have been injected into fissures of the rocks, through which they have found their way toward the present surface of the country. Only in rare cases have they actually passed upward to the surface of the earth toward which they moved; their motion was arrested in the lower levels of the rocks to which the surface has been brought down by the agents of atmospheric decay. The result of their consolidation under the conditions of pressure in which they cooled has caused these originally molten materials to be very compact, a state which is favored also by their chemical composition. This causes the materials to be very solid and elastic. They generally resist decay in such a manner that they often project above the surface, while the softer rocks on either side have been worn down.

The trappean rocks of this country suitable for roadbuilding are plentifully developed in the greater part of New England, except in the northern portion of Maine. In this field they are particularly abundant and of the best quality in the valley of the Connecticut, south of the Vermont and New Hampshire line, and on the shore between Boston and Eastport. In the well-settled portions of New

England there is probably no place which is more than sixty miles distant from extensive bodies of trap of the best quality for road-making.

In the upland district of New Jersey and the neighboring portions of Maryland and Pennsylvania which lie within the belt of old rocks which forms the core of the Appalachian system of mountains, traps also abound; but the valley of the Hudson intermediate between these two fields is, except in its lower parts about the Palisades, inadequately provided with this class of road-making materials. The best of the New Jersey traps, like those of the Connecticut valley, have broken through the Jura-Trias red sandstone, occasionally appearing as the remains of lava sheets which poured forth through the crevices and spread out on the floor of the ancient water-basins of the districts. These sheets, which by the uptilting of the rocks have often been put in very advantageous positions for working, commonly appear as cliff-like escarpments which can be very easily quarried.

South of the Potomac the available trappean rocks are limited to the central mountainous axis of the Blue Ridge and to the Piedmont country on the east. In this Piedmont district wherever the red sandstones, such as are found in Connecticut and New Jersey, exist, there are occasional dikes, so far as observed, of very small size. The writer has noted that these dikes are particularly abundant, though of limited area, in what is commonly known as the Richmond basin, the field whence the coal of that district is obtained. As yet the progress of geological work has not revealed the position of many trap deposits in the Blue Ridge district of the Appalachians south of the James River, but doubtless such will be developed by a careful study of that region.

West of the Appalachian Mountains and thence to the Mississippi dikes are exceedingly rare, two or three only having been described in that realm, none of which as yet afford much promise as sources of road-making materials. In the northern peninsula of Michigan and the neighboring portions of the mainland,—indeed, all about the basin of that great lake,—there are abundant dikes which, though not as yet used as sources of road-making materials, are likely in the future to be extensively drawn upon for the supply of broken stone in the region about our inland seas.

West of the Mississippi there exist, in southern Missouri, Arkansas, the Indian Territory, and at several points in central Texas, sparsely scattered but valuable volcanic rocks of a trappean nature. Of the remainder, the Western plain district, it may be said that dike stones are essentially wanting. In the Rocky Mountains and westward to the Pacific they again abound, together with the kindred lava sheets which often afford nearly as good road materials as do the dike stones. As a whole, North America is less well provided with this group of volcanic rocks, the best for road-making, than is the continent of Europe, where in every region, except on the great northern plain, such rocks are of frequent occurrence and so placed that they can usually be cheaply transported by railways.

GRANITIC ROCKS

Next in value to the trappean rocks we may place those commonly known as granites, including the species of that name, the syenites, and the harder gneisses. The distribution of these materials is, unfortunately, in a general way the same as that of the traps. In common with the last-named group, they have in most cases been subjected to the heating which has permitted them to crystallize, and their chemical constitution insures a certain measure of cementing value. Between the traps and granites about one third of the area of the United States is fairly well provided with road-making stones. These materials may be expected either to lie immediately contiguous to the ways, or to be within a few score miles of railway transportation. There remains, however, the Great Southern Plain, extending from New York to southern Texas, and the broad lands of the Mississippi valley, which are in general too remote from the ancient crystalline rocks to permit the supply of broken stone to be drawn by railway or conveyed by water transportation. In these districts, as in many other as considerable areas, recourse must be had, even in the case of the main ways, to other good but more widely distributed materials. These are in general sedimentary rocks, such as quartzites, limestones, or the iron ores they contain, and the gravels.

QUARTZITES

The rocks known as quartzites, which are products of change which temperature and pressure have brought about in beds originally of a sandy character, are limited in the main to the mountainous districts of the Appalachian and Cordilleran areas and the lesser fields of the Ozarks and the Adirondacks. Beyond the districts above named, though beds termed quartzite are occasionally found, they have generally proved to be too soft or of too little cementing value to make them of much service in road-building.

It should be noted that the rocks termed quartzite vary exceedingly in their value. Certain of them, as, for instance, the deposits in the Berkshire district of Massachu-

4*

setts, though seemingly enduring to the test of eye and hammer, prove in the laboratory or in actual use to be valueless, as the bits do not bind together and are on account of their brittleness soon reduced to a state of sand. It is therefore very necessary for the road-master to be careful in trusting to stones of this description. The laboratory test, made in the manner hereafter to be described (see Chapter V), or the costlier trial on the road, affords the only means of determining the utility of these materials.

LIMESTONES

Among the ordinary stratified or bedded rocks the group of widest diffusion, that alone which is to be looked to in many parts of the Mississippi valley, is the limestones. Rocks of this nature, especially where found in thin layers, with little sign of crystallization, particularly where they contain a small amount of clay, say not more than twenty-five per cent. of that material, often afford tolerable road stones. All the excellent turnpikes of Kentucky, Tennessee, and the southern parts of Indiana and Ohio, in the aggregate many thousand miles of very tolerable ways, have been built of this material. In practice it is found that these thin-bedded limestones, mostly belonging to the Cincinnati group of the Lower Silurian period, wear under a given amount of traffic at least twice as rapidly as the trappean rocks. In some cases, particularly where the slopes are steep, so that the dust readily washes away, or where the road is exposed to strong winds, the rate of wear is about four times as rapid as it would be if the road were covered with the best quality of trap. As the only defect of these limestones is in their hardness and toughness, it may in some cases prove advantageous to cover a foundation, say five inches in thickness, of the broken limestone with a layer of trap not exceeding three inches in depth. The use of this method may be commended in the valleys about the head waters of the Tennessee, in portions of Pennsylvania, in New York, and along the shores of the Great Lakes, where limestones of tolerable quality for foundations abound, and where a top coating of trappean material may be obtained from the ancient rocks on the north shore of those lakes or from the Appalachian district.

In proportion as limestone becomes crystalline, i.e., takes on the character of marble, its value in road-making diminishes, for the reason that the crystalline structure in most cases so far weakens the mass that it is apt readily to pass into the state of powder. As these marbles occur only in districts where better road-making materials are likely to be present, they may not be further mentioned, except to say that their use is commendable for foundation layers, where their fair cementation value makes them tolerably fit for service. So long as the bits are kept from the destructive action of the wheels and feet of the carriages and horses, they lend themselves to the road-master's use. Even where a more resisting top covering of ordinary broken stone cannot be provided, a tolerable road can be made of this material, often very cheaply, by using the waste from quarries, by covering the surface with a coating of ferruginous matter, such as is afforded by the leaner iron ores, or by using a top coating of gravel.

CHERTS

It not infrequently happens that, associated with limestones, flinty layers having a character commonly known as chert are found. Chert varies much in value for use in road-making. Sometimes the material, as is the case with the chert belonging in the lowermost Silurian or Cambrian horizon from Virginia to Georgia, on the western and southern flanks of the old Appalachian field, affords very tolerable road stones, especially when the lower layer of the structure is made of limestone or other material which has a greater binding capacity than the first-named rock. The difference in the value of cherts appears to depend upon the variations in the toughness of the stone. That above noted, particularly the varieties which occur in northern Georgia and the neighboring parts of Alabama and Tennessee, resists the tread of the wheels and feet, and has in most cases sufficient binding power to retain the fragments in place. Although not to be compared in quality with the better traps, the cherts of the southern Appalachian district are commendable and are a precious resource in a region which, because of its clayey character, is in great need of hardened ways, and where higher-grade stone is usually not accessible. Similar, but on the whole less satisfactory, cherts occur here and there in the Devonian rocks of the Ohio valley and are considerably developed in the region about the Upper Mississippi.

South of the glaciated district and within the hill-country of the Appalachians the waste of the chert beds is often found accumulated in great quantities, in the form of what is called taluses, below the cliffs or steep portions of the hillsides. In this position, though the fragments are often somewhat mixed with earth, the material is already quarried and in positions where it can be readily transported to the crusher, or even, with a small amount of picking and hand-breaking of the larger fragments, used in the lower layers of a road-bed. Like accumulations of angular fragments of chert may, throughout the district where the rocks occur, be obtained from the beds of the torrential streams.

CLAY SLATES

Associated with the cherts and limestones, and even, in many regions, where such deposits do not occur, clay slates and other deposits of indurated clayey material occur. Such beds do not afford material which is to be recommended for use on roads. Although the stone is often quite hard to the quarryman's tools, it softens rapidly as it takes in water. It has but slight binding value, and the fragments quickly grind into the state of mud. The life of such material on the road is usually less than one fifth that of good trap. It should only be employed when all other resources for insuring a hardened way are unavailable. The sandstones, except where they have been converted to quartzites or cherts, are also to be passed by in road construction, except where no other means of bettering the way are available. They are practically without binding value, and in almost all cases the fragments quickly go to pieces under the tread of the wheels. Where gravel can be obtained the inferior foundations of sandstone may be made imperfectly to serve by keeping the surface covered to a depth of two or three inches with the gravelly material; but any important main way which lies in a district affording nothing better for road-building than sandstone can often, in the long run, be more economically made by the use of brick pavements, which, though costly in their first construction, are in the course of ten years less expensive than sandstone roads.

GRAVELS

In the road-making materials classed under the head of gravels we have in this country resources which deserve a closer study than they have yet received from the road engineers. In general, gravel may be defined as a mass of small, more or less rounded, fragments of stone which have been broken out and shaped by the action of water or of ice, the bits representing the hardest parts of the bed-rock from which they were removed. As gravels represent the work of agents which bring about the decay of rocks and the transportation of their waste, and as these agents act in exceedingly varied ways, the range of quality of the material is very great.

GLACIAL GRAVELS

Within the glaciated district of this country the surface is generally deeply covered with the waste of the bed-rocks which have been torn from their original place and conveyed for varied distances in the path of the ice movement. The greater part of this debris, commonly that which lies upon the surface of the subjacent rocks, consists of a confused mass made up of clay, sand pebbles, and large boulders, the materials originally commingled with the ice which was left upon the earth at the close of the glacial epoch after the melting of the sheet was completed. More or less mingled with this hodgepodge of debris, which is known to geologists by the name of till, but generally lying upon its surface, we have very extensive accumulations of a sandy and gravelly material resulting from the breaking up of the till by currents of water, which operated during and after the ice lay upon the surface. So general is the distribution of this "washed drift" that it may be said to abound in every considerable valley throughout the glaciated field. In those parts of the area which are known to the writer, including New England and the country to the westward as far as central Wisconsin, it rarely occurs that the washed drift cannot be found within any area of ten miles square.

The washed drift in this and other countries varies much in quality. To a considerable extent it consists of clays and fine sands, but associated with these more abundant deposits in an intimate way there are gravelly accumulations which are often tolerably well fitted for road-building, affording, when skilfully used, a hardened way having endurance which in general may be reckoned next below that of roads formed of granitic rocks. Unfortunately these gravels vary exceedingly in their quality according to the original nature of the pebbles of which they are composed, the sorting action which the water has brought upon them, and the extent to which they have been affected by decay, such as has commonly brought about great changes in the condition of the open-textured glacial waste.

The best of the glacial gravels, those known in New England as "blue gravels," hardly to be classed in this group at all, consist of trappean materials which have been crushed by the impact of the ice and left in or near their seat of origin but little affected by water action. This group of gravels, as is denoted by the term "blue" applied to them, is composed of undecayed rock. So far as the writer has observed, these accumulations are limited to the New England States, occurring in sufficient abundance to have much economic value only in the northeastern portion of Massachusetts. The best gravels of general distribution are those composed mainly of trappean and granitic rocks, with but a small admixture of clay, the less the better. Such gravels abound throughout New England, the Adirondack district, and the Appalachian belt south to the limit of the ice-sheet in that direction. Where, as is particularly the case in the southern portions of New England and in parts of the States to the westward, the gravel is composed mainly of white quartz, because of the smoothness and slight binding power of the quartzose bits the material is of much less value. Wherever these white quartz pebbles constitute more than half the mass it is usually worthless as a covering for roads. It can be made serviceable only by the admixture with it of some binding material, such as iron ore.

The importance of the gravels in this country is the greater as we go westward in the States north of the Ohio River. In that part of the country, because of the soft nature of the horizontal-lying stratified rocks, there is great dearth of road-building stones, which gives the fragmentary material we are considering a relatively large value. Although the drift in the region north of the Ohio is largely composed of the waste from the rocks lying beneath the region which it occupies, a notable part of the covering is composed of fragments borne from north of the Great Lakes, where lie extensive fields of ancient crystalline and eruptive rocks. This transportation appears to have been in part effected by carriage of the bits in the ice itself, but in larger measure was doubtless brought about by the streams of fluid water which ran under the ancient ice-sheet. So effective have been these means of transportation that there are vast accumulations of granitic and other gravelly material from far north of Lake Erie now lying along the banks of the Ohio River. Gravels of the same general nature are associated with the other drift westward for a considerable distance beyond the Mississippi. They are, moreover, to be found along the banks of the rivers draining from the glaciated country for a long way south of the ancient margins of the ice-fields. Thus in western Kentucky, particularly in the region about Paducah, extensive deposits of gravel occur composed in part of Northern waste and in part of materials derived from the banks of the Ohio. These gravels occur there, as they often do elsewhere, above the present flood-plain of the river, their presence at a high level indicating either that the river was of old of much greater volume than at present, or, what is more probable, that the channel has been deepened since the ice time.

STREAM GRAVELS

As the process of separating the gravelly material from the other kinds of glacial drift has been due to the action of free-running streams, we naturally find that it is still going on in the brooks and rivers which drain from the drift area. Thus in addition to the higher-lying gravels, those well above the paths of the existing streams, deposits mostly formed by the subglacial rivers, we have very extensive deposits accumulated in the post-glacial alluvial plains or in the bars along the courses of the existing waterways.

Even where the drift gravels have not, by the process of their accumulation, been brought to a state which fits them for use as road materials, they can often be adapted to the needs by a simple and little costly treatment. The process of screening, best accomplished in rotating drums, may be made to separate the excessively large pebbles as well as the injuriously fine material. Where, as can often be done, the gravel may be lifted by steam-shovels and, in dry weather at least, pass directly through the drums, the cost of the work, based on the rate of delivery from the shovel, including the excavating, is not likely to exceed five or ten cents per ton. Where an objectionably large amount of clay is mingled with the pebbly material it may be necessary to effect its separation by water. This can be accomplished in ordinary " log-washers," consisting essentially of a long trough with a rotating cylinder armed with plates, which stirs the material as water passes through it, the plates being set at such an angle that the gravel is pushed forward toward the point of discharge. With a properly organized equipment the cost of this work per ton of gravel, where the useful part of the mass is as much as one third of the whole, need not exceed, if the work is conducted on the scale of two hundred tons' production per diem, about forty cents per ton. This, however, does not include the cost of excavation, a rate which will still leave the cost of gravel much less than that which is incurred in producing a like amount of broken stone where the material has to be taken from a solid ledge.

BOULDER DEPOSITS

In addition to the abundant stores of gravel which exist in the glaciated districts, or of debris from which available gravel can be separated, the drift deposits commonly include great quantities of boulders. Even where, as in many parts of the Western States, the surface is occupied by a deep coating of soil containing no fragments coarser than sand, there generally exists a boulder-containing layer at no great depth below the surface. Although it occasionally happens that the boulders in the glacial drift are composed of stone having an unsatisfactory quality, all of them are on the average much harder than the mass of the rock whence they were derived. This is due to the fact that the rude treatment to which the fragments borne along in a glacial sheet are subjected speedily wears out the softer materials, leaving only the harder erratics. On this account glacial boulders, if not subjected to decay since the last ice period, commonly afford when crushed a better resource for road-building than anything else except the product of quarries yielding trappean or granitic rocks.

Where the glacial boulders have lain since the ice time exposed to the weather, or where, being of small size, they lie within the zone to which decay has penetrated from the surface, they are often so far decomposed as to be essentially unfit for use in road-making, save it may be as Telford pavement, or as a bottom coating of broken stone which is to be covered with material of a better grade. The variation in the extent to which decay has injuriously affected the surface stone may be judged by simple tests which may be readily applied. After a brief experience, a judicious person with a light sledge-hammer can, by striking the stones which offer themselves to use on the surface of the fields and in the walls, readily determine the state of the masses. If they ring sharply to the blow they may be judged sufficiently sound. If, however, they pulverize under the successive strokes and when broken show evident traces of decay, as by iron stains penetrating the mass, they may be condemned as a source of supply. In many positions the larger boulders and the gravelly material commingled together with more or less sand and clay may be exposed in convenient pit faces, so that the cost of moving the mass either with or without the preliminary breaking up by the use of explosives is not very great. In many such places, better sources of stone being unavailable, it will be profitable to excavate the drift, using the boulders over six inches in diameter in a crusher, separating from the remainder, by the method of washing before noted, the gravelly portions of the mass. By a proper arrangement of the crusher, or perhaps best by the use of two different pieces of such apparatus, all the pebbles above an inch in diameter could be brought into the condition of broken stone.

It is very important that the method of using the glacial waste in the manner above described should be applied in that great portion of the West extending from the eastern border of Ohio to near the Rocky Mountains. Throughout the greater part of that district the soil has the assemblage of qualities characteristic of the so-called prairie lands. These qualities, while they insure an extraordinary fertility and endurance to tillage, make the surface materials the worst possible stuff on which to form a road. The result of these conditions is that the tax on transportation due to ill-conditioned highways is now greater in the rural districts of that part of our land than all the other imposts put together. As all of this country, except the strip next the Great Lakes, may be regarded as impossibly remote for the transportation of road-building stone, we have to look to these generally concealed accumulations of glacial debris for the materials to be used in the betterment of the roads.

CONDITIONS OF GLACIAL DEPOSITS

In seeking available road-building materials in the glacial drift there are certain features in the distribution of that waste which the explorer will do well to note. In the process of retreat of the ice the deposits which it left were accumulated under several quite diverse conditions. One of these produced the till; or commingled coarse and fine materials, which had been churned up into the ice during the time of its motion, came down upon its surface, when the melting occurred, as a broad, irregularly disposed sheet which, with rare exceptions, is to be found in all parts of a glaciated district, save where it has been swept away by streams.

Again, from time to time during the closing stages of the ice age the prevailingly steadfast retreat of the ice was interrupted by pauses or readvances. In these stages there was formed along the margin of the ice-field what is called a frontal moraine, composed of debris shoved forward by the glacier or melted out of it along its front. These moraines are in most cases traceable, where they have not been washed away or buried beneath later accumulations, in the form of a ridge-like heap of waste, which, as we readily note, contains much less clay and sand, and therefore a larger proportion of gravel and boulders, than the sheet-like deposit of till above described. In some cases these moraines are very distinct features in the landscape, appearing, from the number of large boulders which they expose, much like ruined walls of cyclopean masonry. More commonly they are found in the form of slight ridges, which may be covered with fine material, but commonly exhibit here and there projecting boulders. In general it may be said that the moraines afford much better sites for pits from which road materials are to be obtained than the till, and this because of the prevailing absence of clay and sand in the deposits.

Here and there in almost all glaciated districts, espe-

5

cially in the valleys of the greater streams, there may be
found narrow ridges, often of considerable height, and
almost always extending in the direction of the ice movement, as such movement is indicated by the scorings on
the underlying rocks.   These ridges are generally termed
by geologists eskars, and often have a tolerable continuity
for scores of miles at right angles to the ice front.   A
section of them shows generally a gravelly mass, nearly
always free from clay and often containing little sand,
though occasionally there is an abundance of large boulders, which have a prevailing rounded or water-worn
form.   These eskars were doubtless formed in the caves
beneath the ice through which the ancient subglacial
streams found their way.   These under-ice rivers were
much given to changing their position, and as a stream
lost its impetus it was apt to fill its ancient arched way
with debris, which in its time of freest flow would have
been sent forward to the ice front.   At many places in
New England and in New York the writer has observed
these eskars to contain large and useful deposits of gravel,
and also occasionally quantities of boulders well fitted for
crushing as regards their size and hardness.   In the Western States, because of the general coating of deep soil,
these eskars are less easily found; but they exist there, and
should be sought for.

Where the eskars terminate, as they commonly do on a
morainal line, there is almost invariably found, immediately in front of their southern terminations, a delta-like
deposit which, though generally composed in large measure
of sand, frequently contains near the moraine extensive
accumulations of useful gravel and small boulders which
are fit for crushing.

In searching for glacial waste which may be made in

the ways above mentioned to serve in road-making, it is necessary, in the western part of the country at least, to pay little attention to the character of the soil. Information may be had from the banks of streams, where by chance they have cut below the deep coating of fine materials. The existence of any distinct uprise of the surface affords some reason to expect that the coarse glacial waste may be at that point not very deeply hidden. It is probable that the best method of exploration is by any simple form of drill. Even the ordinary post-hole auger may be made to serve the purpose. Where useful material is found it may be worked in open pits, preferably by the use of the steam-shovel; or if, as is the case in some parts of the waste, the superficial coating is more than a score of feet thick, it may be necessary to resort to the ordinary methods of the miner, but even in these cases we may expect, considering the needs of the country, that comparatively cheap road-building materials may be obtained. In general, however, it will probably be better to enter the deposit at the base of a slope at the level of a stream-bed, so that water may be had for washing, if that process is to be used, and natural drainage obtained.

What has been said concerning the sources of supply of road materials in the region of plains north of the Ohio and Missouri rivers, unfortunately does not rest on much practical experience. The writer has personally examined a good deal of the country, and reports of the geologists of the United States Geological Survey afford abundant evidence as to the wide range and much as to the character of the deposits. The cost of excavation and of washing rests upon a large body of experience in the treatment of the iron ores of the Southern States, with which the writer has had much to do. The whole matter clearly deserves a very careful inquiry, as it may afford the solution of the large economic problem presented by the present condition of the roadways in the northern part of the Mississippi valley.

GRAVELS OF THE SOUTHERN STATES

Although characteristic and widely disseminated gravels are limited to the glaciated districts of this country and other similarly affected lands, there are in the region south of the old ice-field large deposits of river gravels, left in the form of terraces often far above the present level of the stream. Little is as yet known of the high-lying river gravels in the southern portion of this country; but enough has been observed to make it certain that these deposits, which often need to be treated by washing in the manner above noted, may afford very valuable road materials, at least in the districts which are not otherwise provided with good road-building stone.

Over a large part of what is commonly called the Southern Plain, that portion of the South which has in general a little indented surface sloping from heights of about one thousand feet toward the sea, there occurs a very widespread deposit composed in large measure of rounded pebbles, small boulders, and gravelly bits more or less mingled with clay and sand. This obscurely bed-like surface accumulation has been termed by Mr. McGee the Lafayette formation. The origin of the mass is not yet determined, but it is probably made up in part of ancient river gravels and in part of the hard portions of a great thickness of rock which has disappeared by the leaching action of the surface waters. Deposits of this nature are tolerably common over a large part of the non-glaciated district of the United States east of the Cordilleras. The sheet often attains a thickness of some scores of feet, and not infrequently has the aspect of a slightly compacted conglomerate or pudding-stone. The fragments in the mass vary much in composition; they are often of quartzite or chert. It not infrequently happens that they contain, to the advantage of the road-builder, considerable quantities of nodular limonite iron ore. It is to be observed that wherever the ordinary unimproved roads pass over a characteristic mass of this Lafayette formation they remain in excellent state throughout the greater part of the year, and are moderately passable even in the worst seasons of alternating rain and frost. It seems likely that at many points it will be profitable to use the material from these deposits, even where it has to be hauled by railways for a considerable distance without any treatment. In other places, as the writer has noted, it will be necessary to submit the mass to a washing process to separate the excess of clay. There can be no question, however, that in a careful study as to the distribution of these Lafayette pebbly deposits there will be found the fittest opportunity for the betterment of many thousand miles of our Southern ways.

PHOSPHATE NODULES

Here and there in the South, particularly in western central Florida, but also in a less determined way in portions of Alabama, Georgia, and perhaps Mississippi, there occur sandy and clayey beds in which are found in variable quantities hard nodular masses of lime phosphate. As is well known, where this phosphate is sufficiently rich for the needs of the manufacturer the material has a high value as a source of fertilization. It is not unlikely, however, that here and there these pebbly beds contain too little

5*

available material to make them useful save as road materials, for which they are very much better fitted than ordinary limestone rock.

At many points in the South, particularly in the States of Georgia and Alabama, there are extensive deposits of limonite iron ore composed of pebbles of that material scattered through clay and sand. It occasionally happens that the admixture of quartz and other pebbles with those of iron ore makes it unprofitable to prepare the material for market at the low price now paid for such ore. Experience has shown that some of these pebbly iron deposits, especially when used with other material such as chert or lime, assist much in hardening a road. This is due to the fact that a portion of the iron, penetrating into the rubble of the way, acts as a cement for the mass. In the absence of other suitable material the consideration of these low-grade iron ores is to be recommended.

It may be said, in closing this account of gravel deposits, that even in districts where no high-lying masses of this nature are to be found the river-beds contain a considerable share of such pebbly matter, which may at times of low water be readily obtained by dredging with scoops after the manner of ordinary steam-shovels operated from a boat. There is probably not a river north of Florida which, in the reaches where the water has a free flow, will be found destitute of such resources for road-building.

SHELL BEDS

At many points along the Atlantic and Gulf coast-line, particularly in the embayed waters, the mud-flats and the low-lying plains covered with marshes are underlaid by deep and wide-reaching fields of ancient oyster-beds.

These accumulations extend as far north as Cape Ann, Massachusetts. Recent excavations on the Charles River flats, Boston, have shown an amazing abundance of such remains, where no oysters have lived for a century or more. In most cases these old oyster-beds preserve their shells in a tolerably firm condition. The cost of excavating the material and of washing away the small amount of sand and mud which lies between the close-packed shells would not be great. If there were sufficient demand for the product, as there might well be, for road-building purposes as well as for making lime, the washed material could probably be furnished at a cost not exceeding fifty cents per ton.

For roads which are not to have heavy traffic oyster-shells afford a tolerable hardening material. At the cost of ordinary limestone, such as is used on the macadamized ways of Kentucky, the material may be regarded as having about the same value as the rock. The use of these shells may be commended in all the seaboard region of the South Atlantic States, where the expense of placing the material on the road does not exceed from a dollar to a dollar and a half a ton. One advantage of using these shells is that there is no occasion for the use of the roller in compacting the mass. The material needs only to be spread evenly, so that the wheels will affect it in a uniform manner. Then, after a few months of use the ruts need to be refilled. After that, with a renewal of the surface coating every three or four years, the depth of the covering in most cases not exceeding two inches, the road may be expected to remain in fair condition. It should be noted, however, that, owing to the rather friable nature of the oyster-shells, ways thus hardened will not resist the tread of heavy carriages. Where there are but few vehicles passing with a greater aggregate weight than about three thousand pounds, such roads, though to be reckoned not as the best, may still prove very serviceable. Some objection has been made to the use of oyster-shells for the reason that the dust from roads to which they have been applied appears to harm the varnish of carriages. The sharp-edged bits are said, also, to damage the rubber tires of bicycles.

PAVING-BRICK CLAYS

In addition to the rock materials, which in what may be called their natural state may be used on roadways, we have to reckon on the group of clays which may be baked into brick suitable for road pavement. The serviceableness of brick to be used as pavement on common roads has been demonstrated by long experience in the Netherlands. Only of late years, however, has this material been brought into use in the United States, so far almost altogether in building high-grade city streets. We therefore as yet know little concerning the distribution of the clays which are fitted for the manufacture of brick suitable for highway constructions. Enough has been ascertained, however, to make it clear that clays fitted for this purpose are widely distributed throughout the section from central New York to the Mississippi River. The general character of the deposits indicates that clays suitable for burning into good paving-brick may be found throughout those portions of the Mississippi valley which are most destitute of stones fit for road-building.

It is required of a clay which is to be used in making paving-brick that it be tolerably resistant to the heat of the kiln, but that it shall, unlike fire-brick, become throughout partly vitrified at the highest baking temperature. The result is a material which, so far as its resistance to the impacts of carriages is concerned, is almost as serviceable as granite blocks. The best results in making brick of the kind now under consideration have been attained by using clays taken from the coal-measures rocks. When first mined these materials are tolerably hard, so that it is necessary to break them up by some method of grinding. Inferior results have been gained from the superficial clays, though the bricks made from them have done very good service.

In those portions of the West where no other materials for hardening the roads can be obtained, resort in time will necessarily be had to paving-brick. Although the treatment of clay necessary in the production of brick is somewhat costly, the use of machinery in the work has of late years very much reduced the expense of manufacturing. Still further, the process of burning by the use of crude petroleum, which can readily be had in the Western fields, or, what may prove to be yet cheaper, by the use of the gases formed in the by-product ovens in which coke is now generally made, may serve still further to diminish the cost of roads of this description. It is not unreasonable to expect that, as there is a large demand for road-paving brick, the cost of ways thus hardened would not very much exceed that incurred in building high-grade roads of trap where the stone has to be hauled by railway, say for the distance of one hundred miles. As yet the greatest expense which has to be met in the construction of brick-paved roads is in the foundations, which it is the custom to make either of plank or of some form of concrete. If by the use of large ordinary tiles or other cheaper material for a foundation the expense of the substructure could be diminished, say about one half, it may be possible to build brick-paved roads even in the present economic condition of many well-developed agricultural districts. It seems, however, to the writer that it will be best to leave this method of construction to attain greater maturity in street pavements than to endeavor at once to apply it in country roads.

Among the questions still to be determined concerning the adaptation of paving-brick to country roads are those as to the proper size and form of the blocks. As yet the shapes which have been well essayed are much like those used for masonry purposes, with a beveled surface at each angle. The proper figure and size of these bricks, as well as the fit method of supporting them and binding them together, need to be the subject of much experiment before it will be well to essay the use of them on roadways. The field is, however, one of much promise.

TRACKWAYS

Where roads are quite unimproved they are often better than where they have been subjected to unreasonable efforts at mending. Thus in the Cape district of Massachusetts, which are the single trackways with thick-set bushes on either side, the teams have been compelled to keep in one line of ruts, except in the places which have been provided for passing. Into the wheel- and foot-ruts the leaves gather, thus serving to bind the sand or gravel together. In the above-named district there are a thousand miles or more of these primitive ways, which are very much better for use than others which have been widened so that there are no barriers to confine the vehicles to one path. In these "improved" roads the sand becomes churned about until the way is uniformly deep. In many instances, unless a road is to be provided with an adequate foundation, it is as well to leave it in the state of a track-way. Such roads have at least a picturesque quality which is never to be found in the unsightly apologies for roads which mark the next stage of advancement.

About forty years ago there was a fancy for plank roads in those parts of this country where suitable timber for constructing them was cheaply obtained. These ways were built in the manner of an ordinary barn floor, except that the boards were laid on stringers and held down by strips placed on the sides without the use of nails. When first built a plank road, if it rests on an earthen support, is excellently adapted to travel. The trouble is that it soon wears out. When it begins to give way the action of the wheels insures its swift destruction; at the end of three or four years it is commonly worn out. A way which it was a delight to drive over thus becomes a very sorry affair.

Among the primitive types of way which may still be found in the swampy parts of this country is the "corduroy" or "laid road." This is formed of small, straight tree-trunks placed side by side across the way. In the better methods of construction these transverse sticks are held in place by heavy timbers arranged lengthwise on the sides of the track. Although a tolerably effective provision against miring, these laid roads are an aggravation to the flesh. Those who have been compelled to traverse them are likely to remember the experience. As some such method of bridging swampy tracts of land has often to be used, it is advisable after the road is laid with the timbers to scarf off the surface of the sticks with a foot-adz. The cost of this process is small, and is more than offset by the much greater endurance of the structure, as it avoids the sharp blows of the wheels which speedily cut the road into bits.

CHAPTER V

METHODS OF TESTING ROAD MATERIALS

Cost of road stone. History of experiments. Methods of testing.
Field-tests of road-making stone. Time required in such tests

As the greater part of the cost of any fairly well-constructed way is incurred in providing the hardening material which is used thereon, it is necessary for the roadmaster to consider the methods by which the fitness of the stones to which he may have access is to be determined. In Massachusetts experience shows that on a road having a width of fifteen feet it is necessary to use between thirty and thirty-five hundred tons of stone per mile, not counting that which may be built into the foundations or applied to the shoulders of the road. The average duration of the material used in the hardened way above the foundations may be taken as about twenty years. Thus the consumption of broken stone per mile of road may be reckoned at about fifteen thousand tons each century.

For the incidental repairs in the periods between each general reconstruction of the way, which has to be undertaken at intervals of about fifteen years, it may be estimated that at least as much more stone would be required, so that the average per annum will be between three and four hundred tons. The above-given rough estimates are on the supposition that the stone used is of the best quality. If it be of lower grade the increase in the amount of stone required may be exceedingly rapid, until at least double the amount of wear and consequent expenditure will be incurred. As the average cost of procuring and applying broken stone, either in original construction or in repairs, may be averaged at about two dollars per ton, the cost in application for repairs being greater than in original construction, it becomes at once evident that endurance, as determined by the quality of the material employed, is a matter of much economic importance. Inattention to this point may double the burden which building roads imposes upon a people.

HISTORY OF EXPERIMENTS

Until systematic experiments in the construction of roads were begun by the French in the early part of this century, the only method of testing the value of stone was by applying the material to the road in some standard manner, and then taking account of the costs of repairs over a series of years. This method, though doubtless the most effective which can be used, is crude, and, as will be readily seen, very costly. To obtain preliminary knowledge concerning the value of stone the French invented a method of placing a given weight, composed of fragments of the stone, in a tight drum so arranged that when rotated the bits were not only rolled over one another, but shifted from end to end of the vessel. After a given number of revolutions of the drum, usually ten thousand, the fine material is gathered and weighed. The ratio which it bears to the original weight of the material is taken as the index of resistance of the rock to such wearing action as it would be subjected to on a road. Although the French method as originally used affords a useful rough test as to one of the qualities which are required in road-building material, it does not give any information concerning sundry other even more important characteristics. It is desirable to know not only how far a stone will wear in rubbing movements, but the extent to which it will break up under the sharp blow inflicted by the feet of horses. Furthermore, as before remarked, a large part of the value of road-making stone depends upon the cementation value of the dust which is produced in the breaking of the material and in the rolling of the fragments together as they are passed over by the roller and by carriage-wheels. Not only the first endurance of the first cementation of these "fines" should be known, but their capacity to reunite when again wetted and dried after being powdered. On this process of recementation largely depends the endurance of any macadamized way.

METHODS OF TESTING

In the experiments on road materials which have been carried on by Mr. L. R. Page, in the laboratory of engineering of the Lawrence Scientific School, principally in connection with the needs of the Massachusetts Highway Commission, the following described method of testing rock which is intended for use on macadamized ways is followed. As in the French experiments, fragments of the material of about the size of those used on the way are placed, to the amount of five kilograms (about eleven pounds), in the rattler, made on the

model used in the French system. The determination
of the resistance to the rubbing and pounding action is
made as in the French method. The general results of this
test are well shown in the table (see Appendix B), which
shows the rate of wear thus obtained on sundry varieties
of rock, mostly from Massachusetts. On the basis afforded
by the experiment above described, the following additional
tests are then made. A portion of the "fines" of the dust
from the rattler is molded into small briquets, cubes having a diameter of three centimeters. These are dried in
a uniform way and are then subjected to the action of a
weight which is dropped upon them at regular intervals,
but with a steadily increasing height of fall. The point
at which the material gives way affords a test as to the
cohesion of the mass. The recementation value is likewise determined by repeating the process, first grinding
the fragments of the briquets, which have previously been
broken, and remolding them, using definite amounts of
water. In the method of testing above described, information is obtained, so far as it has yet proved possible
to secure it through laboratory experiments, concerning
the more important peculiarities of road stones. There
remain, however, some features of less moment which as
yet it has not proved practicable to test in the laboratory.
Among these we may reckon the effect of atmospheric
agents in softening the rock, and especially the influence
of the various acids derived from the dung of animals
which becomes scattered along the way. On the whole,
however, the results of these experiments, so far as they
have been compared with those obtained through practical
experience in road construction, bear out the conclusion
that it is possible to obtain sufficient knowledge, and this
at very small cost, to afford most valuable guidance in the

choice of the species of rock which it is proposed to use in road-building. It is very desirable that this method of testing road stones should be made reasonably free to the public. The federal government could do no better service than to provide such information.

In many cases something concerning the value of the materials chosen to improve roads may be ascertained by observing the effect of the traffic on stone which has been used in filling ruts or cradle-holes. Where the rock is observed to bind well and not to wear with great rapidity, a tolerably good indication that it is fit for service will be thus obtained. Again, it not infrequently happens that the road has worked down upon the underlying rock. Where this proves tolerably resistant to abrasion it is worth while to try a small amount of it in the broken form, laid down in the manner described in Chapter VII. A test of this nature can be made if need be by breaking the stone with the use of hammers in the manner followed until the recent invention of power-crushers. For the purpose of a sound inquiry it is, however, necessary that a considerable strip of road should be thus laid in order that the passing vehicles may take on a normal speed. In general it may be said that such an experiment is not likely to be thoroughly satisfactory unless it includes level ground and as steep a grade as is likely to be called for in the larger use of the material. In most cases these conditions of practical testing cannot be obtained with less than one half a mile in length of road.

A test of some value could probably be made by building small strips of road with the broken stone which is to be submitted to inquiry at various points on a considerable length of road, choosing the level and the steeper parts of the way as the sites.   If care were taken to have the pieces laid in such condition that the vehicles would pass .over the experimental pavement at the speed they would be likely to have on a thoroughly improved road, there is no reason why this method of testing should not be satisfactory.

TIME REQUIRED IN SUCH TESTS

The time required for the essays above mentioned in laying experimental bits of road is unfortunately considerable.   No final verdict as to the value of the material can well be obtained with less than five years of trial on a road which is considerably frequented by teams. Where, as on many country roads, light vehicles only are used, and the traffic does not involve the passing of more than thirty or forty vehicles each way in twenty-four hours, the test to be sufficient would probably have to be extended for at least eight years.   It is this time element in such experiments which makes it so desirable to have the stone subjected to laboratory tests.

CHAPTER VI

THE GOVERNMENTAL RELATIONS OF ROADS

Difficulties of problem. American systems of road management. Need of control by States. System of control by private corporations. Methods of control by the State. System of Massachusetts. Methods of building State roads. Methods of supervising. Relation of electric roads to highways

THE difficulties which have been encountered in maintaining a well-constructed system of public ways in any locally governed community have always been very great. It is a noteworthy fact that the early successes in road-making were altogether limited to the Roman empire, in which there was a strong central government controlling all matters relating to public ways, in which the people of the communities in which the roads lay contributed only their enforced labor, the plans for the location of the route and for its construction being determined by authorities over which the provincial folk had no control. With the downfall of the Roman empire, even where the economic estate of the locality gained by the change, we find that all system in road-building ceased to exist.

The revival of the art of highway construction came when the government of France had become thoroughly centralized so that it could play a strong part in the administration of the provinces.   The bettered conditions of England, which began somewhat after the improvement in France, were likewise due to a considerable centralization of authority, the people of the vicinage having little to say as to the construction and maintenance of the principal ways.   The system of toll roads under the authority of charters has hitherto proved the principal resource of decentralized governments in improving the ways of the people.   Beginning in feudal times as a convenient means of levying taxes, the toll system was gradually in more modern days so modified that these ways were handed over to corporations which, while they were at times called upon to pay some share of their income to the coffers of the state, were required before dividing profits to maintain them in good passable condition.   Long after the sovereigns ceased to have any actual control over the roads in the way of toll-taking they were still, and to this day in England are, known as the king's highways.   The term continued to be used even in this country long after the separation from the mother-country.   It may be heard occasionally even to this day.

AMERICAN SYSTEMS OF ROAD MANAGEMENT

When the English settlements were made in North America the idea of a central control of road systems had not been redeveloped after its loss in the middle ages. The only method of effecting improvement was that by means of corporations which were allowed to tax those who used the ways.   Although there have been occasional interferences on the part of the federal government, and of State governments looking to the improvement of particularly important ways, it was not until within a decade that the idea of managing the principal highways of any commonwealth by a State board was brought into prominence.   Up to the present time it may be said that our American system, or rather lack of it, in the management of roads has rested upon the action of local authorities, those of counties or of towns.   Here and there in wealthy communities public spirit has led to a large, though almost uniformly wasteful, method of bettering the ways, with the result that in a number of counties which could be named, in all not including perhaps more than one one-hundredth part of our population, the highways have been brought into a tolerably satisfactory state.   Experience, however, shows that, even with people as patriotic in relation to the interests of their neighborhood as Americans generally are, it is impossible to develop a well-considered plan of roads.   Even where these exist within the limits of a small community they are never properly related to those of the adjacent administrations. There is no sense whatever of the commonwealth in their planning or administration.

NEED OF CONTROL BY STATES

There are sufficient reasons in the nature of man why it is impossible in any state to provide a fit system of highways by the action of boards which have only local authority, and which are necessarily swayed by purely local, if not individual, interests.   As before remarked, no good system of roads has ever been developed without a large authority lodged in the hands of some central administration.   Under any other system we may expect at best occasional good roads, which will serve only the needs of

A ROAD NEAR TRYON CITY, NORTH CAROLINA.

those who pay for them, and will have no reference to the poorer or less enterprising communities which may lie on either hand.   It is therefore evident that the alternative in the matter of bettered ways is between a system of what may be termed state roads, or roads which, as regards their location and maintenance, are under the control of some central authority and the toll system.

SYSTEM OF CONTROL BY PRIVATE CORPORATIONS

The consideration of the toll system may be shortly made by a slight study as to the effect which it has on the development of a community, and the tolerance by that community of the method.   The best illustration which is to be had of this method is to be found in the commonwealth of Kentucky, where practically all the important highways, to the extent of several thousand miles in length, are held by corporations.   In the beginning of the development of Kentucky the toll system proved very advantageous.   The great highway connecting that commonwealth with West Virginia through Cumberland Gap was built and maintained, and is still owned, by the Wilderness Turnpike Company.   So long as the travel over this road was large in amount it was kept in excellent order. being at one time in the early part of this century by far the best-conditioned highway of considerable extent south of the Potomac; it was claimed, indeed, that it was the best in the United States.   As the traffic over the road has fallen away, the revenues from it have been too limited even to keep the road in condition for travel, and yet the system of toll-gates has been maintained.   The writer has paid a toll of two dollars in passing one of the gates, which are about seventy miles apart, although he found it necessary to take tools in the wagon in order that he might be able to repair portions of the way which were really impassable. In the greater part of Kentucky, particularly in the limestone districts, the toll-roads are maintained in a very good state of repair. The tax on a two-horse farm-wagon for a round journey is about three cents a mile. On pleasure carriages, as they are classed, it is often at a much higher rate. In general, where these roads have not been paralleled by railways, they have proved exceedingly profitable investments to the stock-holders, many of them at the present time paying dividends of from ten to twenty per cent. per annum. Under these conditions the roads are, as might be expected, maintained in an extremely costly manner. The stone is broken by hand, is applied without the use of rollers, the bridges are very poor, and no effort whatever is made to better the grades of the original construction, which in almost all cases are extremely bad. The effect of all this is to put a very heavy tax upon the farming class. The roads are mostly held by the capitalists of the towns, and are administered solely with reference to dividends. In this way the free intercourse of the people is obstructed; the country folk of the poorer sort often cannot afford to make any journeys save those which are certain to bring them a good money return. Even the attendance on the schools is hindered by the fact that children often cannot be transported to these institutions, which in a sparsely settled country such as Kentucky are often remote from the dwellings. Thus, although in opening a country to settlement the system of toll-roads is often advantageous, in that it may persuade capitalists to make improvements which the frontiersman cannot afford, the eventual influence is to retard, or even to completely arrest, the full development of the economic life of those who dwell on the land.

The general condemnation of the toll-road system is seen in the fact that such roads have been completely abolished in the greater part of the prosperous eastern sections of this country, as well as in the Old World. Any extension of the method would be distinctly against the best interests of the people, save, as before remarked, in the frontier districts, where in all cases the eventual freeing of the ways should be provided for. One of these methods is by requiring that a portion of the money received for tolls should be used as a sinking-fund to retire the stock of the corporation.

METHODS OF CONTROL BY THE STATE

Dismissing as we must the idea of bettering our American highways by toll-roads, and abandoning, for the reason that it has been condemned by experience, the plan of controlling these ways by local boards, we are forced to consider the action of central authorities—those of the State or that of the federal government. Although there have been several propositions brought before Congress for action on the part of the federal government looking to the institution of a system of national ways which should afford substantial relief to all parts of the country, it is futile to expect a betterment from this source of authority. The theory, and the mechanism alike, of our federal government is opposed to any such local work as would be necessary to provide a proper system of intercommunication throughout all parts of the land. It would not be unreasonable, if highways were used as grand routes for marching armies across the wide country, to expect a construction of certain national roads; but in the present condition of our transportation system the ordinary highways are of small account for long journeys; rarely, indeed, even in the present incomplete state of our railway system, does a wagon convey its load for the distance of more than thirty miles. Thus our highways have become matters of local interest, and are therefore properly outside of the consideration of the federal government.

It is otherwise within the limits of our separate States. Those units of government are, with the possible exception of Texas, sufficiently limited as regards area and sufficiently united in purpose to make it fit that the authorities of each commonwealth should undertake to control the main highways within its bounds. The method of representation in the State legislatures is such as is likely to secure proper attention, on the part of a State board having the matter of highways in hand, to each of the local divisions under its administration. Moreover, as our State governments are usually supported by direct taxation, it is the interest of the central authority to develop the economic resources of every portion of the area.

To show the need of interference on the part of the commonwealth in its road system, it may be well to note the results of two and a half centuries of experience with the highway problem in Massachusetts. Although this State is, for its area, perhaps the wealthiest in the Union, and has been characterized by a large measure of public spirit, its highways are, with rare exceptions, in a very unsatisfactory state. Four years ago, after a long-continued debate and many experiments in the management of the highways, the legislature provided for the establishment of a commission to take account as to the condition of the country roads. The report of this commission, of

ERRATA

Page 3, tenth line from foot, for "until," read "about."

Page 13, fourth line, for "As will be seen by the diagrams," read "As shown by diagrams 4 and 8."

Page 37, eleventh line from foot, for "compared," read "computed."

Page 52, fifth line from foot, after "tons," read "per mile."

which the writer was a member, showed that by far the
greater number of the towns in the commonwealth were
very ill provided with means of communication leading
either to adjacent towns or to the main railways.   Some
of these remoter towns were, measured in cost of transportation of a ton of freight to Boston, further removed from
that city than Omaha, Nebraska.   In making the survey
of the roads of the commonwealth the commission followed
the plan of sending photographers along certain of the
more important through routes with directions to take
views of the road at intervals of a mile.   The result of
this presentation was very impressive.   It showed that
for many months in the year long stretches of these main
ways were unfit for use by loaded vehicles.   The testimony
taken at many hearings held by the commission showed
very clearly that the tax upon the industries of the commonwealth due to bad roads probably exceeded all the
other imposts upon the agricultural lands.   It became indeed evident that the abandonment of farms, so common
in many portions of the State, was largely to be explained
by the bad condition of the roads, and the consequent burden upon the economies of the people.   The result of
this presentation, made in the report of the commission
for 1892, was the passage of a bill which provided a means
whereby a commission to be appointed under the act
should accept as State roads the more important rural
ways of the commonwealth.

SYSTEM OF MASSACHUSETTS

Although Massachusetts began the new system of State
highways some years after the State of New Jersey undertook a plan for the general betterment of her roads, the

plan of the first-named commonwealth is one that provides more distinctly for State interference than does that of the more southern community. In fact, the Massachusetts system has become, in a somewhat general way, accepted as the most promising experiment in the direction of State interference with road construction. It will be well, therefore, briefly to set forth the method which is pursued in that State.

In all parts of this country the desire of the people to retain control of their local affairs is strong. That state of mind is indeed the safest index of a true democracy. The town organization of Massachusetts gives to the three hundred and fifty small municipalities a very large share in the management of their own affairs. This system has existed since the early colonial history of the common-wealth. On this account the law providing for the insti-tution of a State Board of Highway Commissioners (see Appendix A) made it impossible for that commission to consider the question of accepting as a State road any way which had not been petitioned for either by the selectmen of a town, the mayor and aldermen of a city, or the county commissioners of some one of the fourteen counties in the State. Whenever the commission, after receiving a petition from any of the authorities above mentioned, accompanied with a plan and profile of the proposed way, adjudge that the public convenience and necessity require that the State should take possession of the way, they file a plan of the same, together with a dec-laration of the appropriation of the road, in the clerk's office of the county in which the road lies. After thus taking control of a way it has in all cases been found necessary to reconstruct it. The mode in which this work is done depends upon the needs of the traffic which is to be accommodated by the way. Of the seventy roads which have been accepted by the commission, all save two have been rebuilt with the use of broken stone, and where necessary of Telford foundations. The exceptions include one road hardened with gravel, and one which is to be built with granite-block pavements.

METHODS OF BUILDING STATE ROADS IN MASSACHUSETTS

In substantially all cases it has been found necessary to reform the grades of the ways which have been acquired by the State, and in some instances it has appeared necessary to change the original position of the road, so as to diminish the steepness of the grades or to better the natural foundations. In general the roads thus constructed have a width of location of from forty to sixty feet. Where, as was found to be the case in most instances, the right of way was less than forty feet in width, it has almost always appeared necessary to insist on the municipality acquiring such land as was necessary to give at least fifty feet for the possible future extensions in the width of the pavement and sidewalks. Where it has seemed likely that the location would have to afford a place for an electric railway, effort has always been made to obtain a width of not less than sixty feet, it being evident that a double-track tramroad could not find a place in a location of less width without cramping the wheelway or the sidewalks.

After much consideration the commissioners determined to limit in general the width of the worked road to about twenty-six feet, three feet on each side being allowed for the gutters. Within these limits, three feet on each side are allowed for the shoulders of the way and fifteen feet for the hardened part of the road, that which is paved with the usual depth of broken stone, though the shoulders are also, in most cases, in a measure hardened either by a coating of gravel about four inches in thickness or by a similar coating of broken stone of small size, where gravel cannot be obtained.

In the beginning of the task of construction the people of the country seriously objected to the narrowness of the hardened way, claiming that it was insufficient for the needs of travel on any but the least frequented roads. The commission, however, had made its determination on the basis of a careful measurement as to the width of the traveled portion of a great many much-used country roads. The experience of two years has affirmed the conclusion that fifteen feet is usually a sufficient width for a wheelway, provided there is an additional marginal extension of the road in the way of shoulders, which are covered, as noted, with a resistant coating, upon which carriages may occasionally turn out without any risk of seriously rutting the surface. So satisfactory has this experience been that one main road in Dukes County was last year built with a width of twelve feet of broken stone. This way, which may be termed of the third order of importance, having a daily passage each way on the average of from forty to fifty vehicles, has proved sufficiently wide for the needs. It should be noted, however, that though the hardened portion of the road is narrowed, the distance between the outer angles of the shoulders is kept the same, so that if at the time of general repairs it should appear desirable to widen the hardened section, it can be done without much cost.

As will be seen from the specifications (see Appendix A), as well as from the law constituting the commission, provision is made for two modes of contracting for the construction of State highways in Massachusetts. On the basis of an estimate of costs made by the commission, the town authorities or those of the cities may make direct and non-competitive contracts for building the road. If, after thirty days from the time of receiving notice that the commissioners are ready to make such contracts, the municipal authorities neglect or decline the arrangement, the work is advertised and let to the lowest bidder, with the limitation that the contract has to be approved by the governor and council. Under either of these conditions of contract, the commissioners place a resident engineer on the work, who, following carefully arranged printed instructions, sees that the requirements are exactly fulfilled. He, moreover, obtains data which will enable the commission exactly to reckon the cost of the undertaking; not only in general, but with reference to each item in great detail, so that information may be had which will serve to guide in making future estimates. The result of this system has been that the towns, by far the greater part of the number of those that have taken contracts, have accomplished their task with insignificant profit or loss. In a few cases, where political influence or personal favor has entered into the work, the usual result of such illegitimate motives has been found in the more or less considerable excess of expenditures.

The advantages arising from contracting with the towns are numerous and go far to countervail a certain increased cost, probably not more than from five to ten per cent., which is brought about by this method of doing the work. In the first place, in practically all cases the task of building the road is altogether done by the citizens of the town, so that, guided by the resident engineer, the people learn how a good road should be built. It often happens that from thirty to fifty persons acquire more or less of this knowledge, and thereby become critics of no mean order as to the conditions of highway construction in their own district. In several instances it has come about that no sooner was the town done with the work on a State road than it has turned at once to the construction of its local ways on the same general plan. Moreover, with a considerable assured contract in hand, many towns have been induced to purchase road-building machinery, particularly crushers and rollers. The possession of such apparatus leads naturally to a better system of ways in the community which owns them. Not the least of the advantages of the plan of contracting with the towns, one which is very much appreciated by these social units, is that it avoids the evils arising from private contractors bringing alien laborers within their limits. These imported workmen not only deprive the citizens of employment which they value, but it often comes about that the least worthy of the gang are left in the community, to become undesirable citizens or perhaps paupers. Of the ninety-four or more contracts made by the Massachusetts Highway Commission only five were awarded to private contractors.

At first sight it may seem impossible, under a system which does not permit a commission to determine of its own instance the position of local ways, to connect these roads in such manner that they will serve not only the interests of the community which may petition for them, but also those of the commonwealth as a whole. Experience, however, has shown that even where the petitions come not from the county commissioners, but altogether from the municipal authorities, it is by no means difficult, by preliminary conferences with the people of the community, to have the applications so determined that an extended system can be arranged as effectively as though the separate pieces of road were selected by the State board. It is to be observed that this board is not compelled to accept any way save where, in its opinion, the public convenience and necessity demand its taking over by the State. This option has, together with the good sense of the people concerned, made it easy in all cases to arrange for benefiting the petitioners in such a manner that the undertaking shall relate to others in the same field.

The first enactment concerning the commission provided that two or more cities or towns should join in the same petition. Objections to this requirement arose from the natural jealousies of neighbors, and on this account the law has been changed so that a single municipality is competent to petition. Experience has shown that the commission can arrange more effectively than the law for community of interests as regards the location of the intertown ways.

It will be observed that under the Massachusetts system the commonwealth bears three fourths of the expense of constructing and maintaining the State ways. One fourth of the total is taxed upon the counties, with a provision for repayment distributed over a term of years. It has been proposed that a portion of the charge be laid upon the city or town in which the work lies. Although there appears at first sight to be a measure of justice in this arrangement, it has the disadvantage that it will heavily tax some of the communities which most need the benefits to be derived from the State-road system. Many of the towns in Massachusetts have a total valuation of less than half a million dollars. Several of these poorer communities require the construction of from three to five miles of road. Any plan which would assess a material portion of the cost of such ways on the towns would be hurtful in that it would for many years bring the tax levy above the endurable point, thus leading to further pauperizing of the communities which it is designed in an economic way to better.

The aim of the Massachusetts commission has been to distribute its constructions over the State with reference to various classes of needs. In the first place, it has endeavored to better the important roads which are already of much value to the industries of the commonwealth. In many cases a main way which is costly to maintain and which principally concerns two rich communities passes through a poor town, which has unfairly to bear the cost of repairs, with the result that the road generally falls into very ill condition. In the next place, endeavor has been made to better the situation of certain towns which, owing to their distance from the main centers as well as to the neglected state of the roads connecting with their markets, are incapable of developing industries and are often in course of decay. The aim is to provide these places with at least one first-class main road giving access to markets. The effect of these improvements, though they are but in their inception, has already begun to be felt. A number of the hill towns of the Berkshire district are now being approached by State roads. There has already been in anticipation of the benefit a rise in the value of their lands, and the people are looking forward to utilizing the water-power and the timber resources which abound in that part of the country.

METHODS OF SUPERVISING ROADS

As yet, owing to the fact that none of the Massachusetts roads have been two years in use, the problem as to the method of repairing them has not been fully determined. It seems likely, however, that the best plan will be to commit this matter to the town authorities, they undertaking, on the order of the chief engineer, to keep a watch upon the ways, which will furthermore be inspected each month by an officer of the board.   This system appears to be desirable for the same reason that warrants the endeavor to have the construction effected by the local authorities. Where the officers of a municipality become accustomed to keep one or more of their main ways in perfect repair, they will be more likely to deal in a like careful manner with those for which they alone are responsible. Thus the educative effect of this system promises to be not the least portion of its advantages.

It will be observed that the plan of turning over the care of State highways to the municipal authorities, under the supervision of the State board, is a distinct departure from the methods pursued in France and other European countries.   In those lands the roads are altogether managed by the central authority.   It will probably be found in our American conditions that the concession to local pride and prejudice will have to be made, and that by making it a decided gain will be secured which could be obtained in no other way.

It will be well to say that, while the law does not permit any State road to be torn up without the consent of the commission, the local authorities have a right to take such action as may seem necessary to meet the needs aris-

7

ing from accidents. Moreover, under the laws which at present exist, the right of granting concessions, as for the passage of a street-railway, appears still to reside in the municipalities, and, though the roads could not be disturbed for any such purpose without the consent of the commission, it seems possible that the courts might force that consent. It is an open question whether it may not be best to extend the control of the State board so that placing street-railways within the limits of the location might be at its discretion.

RELATION OF ELECTRIC ROADS TO HIGHWAYS

The recent extension of electric railways into the country districts, a feature which promises to be a revolutionary innovation in our methods of transportation, is likely to make certain necessary changes in our highway system. In the first place, as before noted, the conjunction of a tramway and a public road requires a wider location. Where, as is often the case, the railway needs to be provided with double tracks, the Massachusetts commission has insisted that the location or right of way of the road shall be wide enough to admit placing these tracks in the middle, with the improved road on one or both sides of the rails. On the supposition that there are to be sidewalks having a width of not less than seven feet, gutters on each side of an aggregate width of six feet, and that the hardened way will eventually be in two sections, each fifteen feet wide, on each side of the strip occupied by the rails, the location should have a width of not less than sixty-six feet. Where there is no reasonable expectation of more than a single-track tramway, it is clearly best that it should be placed on one side of the road, not being allowed to cross from side to side as is often the case. In this arrangement the minimum width of the location should be fifty feet—it had best be sixty feet. It will be observed that the estimates for width as above given for streets which are to be occupied by iron ways are the least which will permit a convenient development of the two methods of travel beside each other. In the opinion of the writer, a street which is to have a double-track railway should have a width of not less than seventy-five feet. This will permit the construction of sufficiently wide sidewalks and also of a row of trees, with or without a narrow grass strip on each side of the double-track railway. Such a provision is most desirable, not only for the general effect of the way, but also to prevent the excessive development of dust which is apt to take place where there is a broad ungrassed tract of land such as would otherwise exist.

It has been objected that, as the electric roads are in the control of private corporations, the exacting of toll for their services justifies demands that they be required to provide their own locations. Granting the equity of this claim, it may be said that these roads are public conveniences of a high order of value; that they will not serve the people so well in any other position as on the main highways, and, furthermore, that their use for passengers, and in time for freight as well, will diminish the traffic tax on the ordinary roads which they parallel. These considerations seemed sufficient to warrant the authorities in endeavoring at once to provide ample width of location for all the roads which these tramways are likely to traverse.

Experience has shown that, under the Massachusetts laws, a State board is likely to be fairly inundated with petitions for the acceptance of ways by the State. In almost all cases the effort is natural for each community to get rid of the large impost occasioned by the most important and worst-conditioned of its roads. It has been the policy of the Massachusetts commission to accept those ways which were most important and most defective, and in proceeding with the reconstruction of any of these ways to take at once what may be called the critical portions of the line; that is, if a petition covered five miles of road, and there was a section one mile in length which, from its state as regarded grades or foundation, was the most difficult to build and maintain, work was begun on that portion. The result is necessarily that the pieces of road, numbering in all about ninety and amounting to a total of about eighty miles, which have been built have cost, including the proportion of office and other expenses as well as construction, a total of rather more than seven hundred thousand dollars. The exact sum is not at present statable for the reason that sundry of the undertakings require a certain amount of expenditure before they will be completed. In other words, the average cost of these worst pieces of way in the commonwealth has been about nine thousand dollars a mile. As will be noted in another chapter, this expenditure, made upon peculiarly bad ground, is probably at least one third greater than that which will be required per mile to provide high-grade rural ways.

CHAPTER VII

THE RELATION OF PUBLIC WAYS TO THE ORNAMENTATION OF A COUNTRY

Esthetic conditions of roads. Roadside trees. Roadside plantations.
Roadside parks. Water-supply on roads. Bridges

IT is a good feature of our day that the people of this
country, so long neglectful of all considerations of
beauty in the landscape about them, have not only become
interested in that element of culture, but are willing to
make considerable sacrifices in order to adorn the land
about their dwelling-places. Inasmuch, therefore, as roads
are important elements in a landscape, serving greatly to
elevate or to debase the view, it seems fit to give some attention to the esthetic quality of roads.[1]

In general it may be said that in proportion as a road
is so laid out and built as properly to serve the needs of
the country it traverses, it fits harmoniously with its natural features. Thus a way which is accommodated to the
irregularities of the surface, and which is evidently so
placed as to afford an easy route into the land, almost always if well built is a gracious addition to the prospect.

[1] Mr. H. Langford Warren, now professor of architecture in the
Lawrence Scientific School of Harvard University, has written an
excellent short paper on this subject, entitled, "A Plea for Esthetic
Considerations in Building Roads." (See "Pavements and Roads,"
E. G. Love, editor, New York, 1890.)

If, on the other hand. it is forced across the field of view,
climbing the hills abruptly and in other ways disobeying
the injunctions of nature, the effect may be in a high
measure offensive. With roads, as with the other features
of human handiwork. they may enter into the nature about
them. becoming agreeable marks of the hand of man, or
they may show him to be an offender against the order
of the world.

ESTHETIC CONDITIONS OF ROADS

The first prescription for the construction of a road
which is to add to and not detract from the beauty of a
vista is that it shall not be obtrusive. This element of
offense may be avoided by keeping the way on easy grades,
so that it may not appear, as some roads do, as if hung
against the hills. The second point is to avoid an unnecessary width of the traveled way. A road having a pavement not more than fifteen feet wide appears from any
point of view as a thread in the landscape, while if allowed
to come into the condition of many of our country roads,
where a space fifty or sixty feet in width is plowed by
irregularly disposed ruts, the impression made upon the
eye is disproportionately great.

In practice, where beauty and utility as well are sought,
the shoulders of the road should be kept in grass. Where
the gutters are not paved in the manner described in the
chapters on construction, it is best that they also should
be kept in a close sod. The slopes left in grading should
be brought to a declivity of not more than about thirty
degrees, sown in grass or planted with some coarser vegetation. For this purpose varieties of coniferous trees are
often the best adapted.

ROADSIDE TREES

It is a good custom in all countries where roads are well cared for to plant trees along their margin. Usually these plantations are made in single rows of equally spaced plants just outside of the gutters. This arrangement has its convenience, and where the object is to shade the way the plan commends itself. In general, however, experience is against any form of tree-planting wherever the road, unless it be a mere single-track lane, has not been provided with some kind of firm pavement. In summer on "dirt roads" the tendency of anything like a complete shade is to keep the way much too wet. The trees if they are strong-growing are apt to extend their roots under the gutters, and even beneath the roadway, in such a manner that in course of time they tend to disrupt the structure. In the winter season the effect on snow is also damaging. If there be open country on either side of the plantations, the diminution in the speed of the wind when it passes through the trees is often sufficient to bring about the accumulation of snow-drifts. In some parts of this country the breaking out of snow-drifts is one of the most considerable elements in the cost of keeping highways in condition for traveling. There is furthermore an artificial quality given by regular lines of trees, which, however, may be compensated for by their stateliness, or, where they have a columnar growth such as characterizes the Lombardy poplar, by a certain architectural effect.

The greatest measure of adornment of a road is accomplished by systematic plantations of groups of trees on either side of a traveled way, the species being varied and the outline of the plantations toward the road broken so as to promote pleasing vistas. Such work should, if possible, be planned by a landscape architect, or at least by a person who is familiar with the expression given by trees in their adult growth. The result of such planting is to afford, even within a narrow belt, the effect which is obtained in the higher class of parkways. If the road is alined so as to avoid offensively long, straight vistas, the charm thus won is well worth the trifling expense which it entails.

The cost of systematically planting, under ordinary conditions, the borders of a highway, including the price of trees, the labor of proper planting, and the care which is demanded for the plantations until they are able to shift for themselves, need not exceed from three to six hundred dollars to the mile. There is probably no other means of rural adornment where so satisfactory and enduring a result can be obtained in a short time and at so little expense.

The species of trees to be used in roadside plantations need to be determined in relation both to soil and climate. Probably for the reason that the early plantations of elms in rows along the New England village streets proved successful, the growth of the species being rapid and their forms graceful, the elm has become the favored street tree in all the Northern and Eastern towns of this country. Of late years, however, the species of this group have been so much preyed upon by varieties of insects which are now firmly established that they are no longer to be recommended. Among those to be suggested as useful are the maple, particularly the water species, though the ordinary rock-maple is also serviceable, the red oak, and the sycamores, which in damp ground grow with fair rapidity and have very beautiful trunks and branches.

Although some objection has been made to the nut-bearing trees, for the reason that they tend to litter the way and to incite boys to use them roughly, the writer is disposed to recommend this group as among the most satisfactory for road trees where the planting is to be in single rows. Of these nut-trees the hickories and walnuts are clearly the best for ordinary use. The various species grow well on most soils, except the more arid, and they are remarkably exempt from insect plagues. They are, moreover, little inclined to send their roots horizontally to a great distance. Beeches are to be commended, except for their very slow growth and the fact that their roots run near the surface of the ground. In many parts of Europe it is the custom to plant fruit-trees next the way in regular lines. Of these the cherries are the best because of their rather dense foliage. In the Old World it is well understood that the fruit of the wayside trees belongs to the proprietor of the land, and it is safe from marauding. In the present condition of the sense of property in this country, it is not likely that the fruit would be respected by the wayfarers.

The planting of nut-trees may be commended for the reason that there is now a considerable market for walnuts and the shagbark hickories, as well as certain other species such as the pecans, which will grow in the region as far north as the Ohio River. Even if the product of these trees should be regarded as public property, some profit would be had from gathering it. At any rate, the crop would afford pleasure to the young people.

ROADSIDE PLANTATIONS

An opportunity of attaining distinction awaits some enterprising community which will undertake to make its roadsides in the springtime an exhibition place of our flowering trees. Those who have traveled in Japan descant much on the charm or even the splendor which the waysides exhibit in the time when the cherries blossom. The variety of native and foreign trees, including the catalpas, the so-called Florida dogwood, the redbuds, etc., is such that, with a little care in plantations, a marvelous floral effect could be produced. A similar floral exhibition in the autumn is easily winnable by the use of the various species of composite flowers, asters, goldenrods, and other species, which readily lend themselves to plantations.

ROADSIDE PARKS

Now that the advantage of public reservations, such as parks, is much considered by our people, it appears desirable to organize such parks or commons with reference to the main highways. On almost any road having a length of three miles or more it is possible, in New England at least, to select one or more attractive bits of ground which may be devoted to this use. These reservations need not be of considerable area in order to obtain effective results. It often occurs that a strip of land next a river or a lake which is skirted by the road, or a bit of picturesque rocky ground, can be obtained by gift or at a low money cost for the reason that the place has no agricultural value. Although it is desirable that these dedications to public use be cared for, it is often better that they should be left in their simple wilderness state rather than be made the seats of elaborate ornamentation. A study of Massachusetts roads, which the writer has made with some care, indicates that a thousand reservations of the nature here indicated could be obtained by purchase at a fair money value, at a total cost of less than one hundred thousand dollars, the average area being not over five acres. A probable total length of the roads in Massachusetts which are to be taken over by the commonwealth is about two thousand miles. Thus the system above proposed, if completely applied, would give park bits at average intervals of about two miles.

Some years ago the legislature of Massachusetts constituted by enactment a Board of Trustees of Public Reservations. The object of this organization is to afford a safe corporate body empowered to hold bits of land which may from time to time be dedicated to public use. This board now possesses a number of such reservations. It gives promise of affording an excellent means whereby lands may be held safe from the temptation which would beset municipalities to part with such holdings at the solicitation of persons of much local influence. The transfer of roadside park places to similar commissions in other States is to be commended as a measure of safety against encroachments such as have served to destroy many of the original commons in Massachusetts and other States, and which in Great Britain have lost to the people more than three fourths of the lands which were public property three centuries ago.

In many of our rural communities, under diverse names, there exist village betterment associations which might well coöperate with those who are engaged in constructing highways in the task of adding to their beauty and to their value as public places.

WATER-SUPPLY ON ROADS

An important adjunct to any well-constructed road is the system of watering-places for the refreshment of man and beast. Here and there in New England and other hilly parts of the eastern United States the people **have** availed themselves of the chance to provide such a water-supply from high-lying springs.   In other portions of the country where streams cross the road an arrangement is made by which the horses may be turned from the main way into the water.   In some of our States an annual allowance is made in the way of rebate of taxation to those who may provide suitable watering-troughs and keep them in repair.   In general, however, this provision for man and beast is neglected.   The writer has passed over many strips of ten miles of highway on which it was impossible to obtain water for horses without begging it from the farms.   There should, indeed, be well-enforced laws requiring that no main way should be without arrangements for watering animals at intervals of not more than four miles.

Where the character of the country is such that water can be brought in pipes to the road from points not more than five or six hundred feet away, this supply by gravitation is in practically all cases the cheapest and best that can be obtained.   Pumps are likely to get out of order; they demand constant attention, as do also the wells from which they draw their supply.   Where wells have to be resorted to, as is the case in most plain lands, the use of driven pipes with small windmills is to be commended. If the well has to be sunk to the depth of thirty feet or more, the pipe method where applicable is almost always the best.   A small windmill, such as can be provided with a sufficient tower for less than one hundred dollars, can, with a little attention, be made to serve the needs.

In most cases the watering-troughs of this country are too large for the flow of water which passes through them. The result is that the saliva of animals remains in the basin as a source of contamination and disease.   The

most satisfactory troughs are those made of single blocks
of stone, with a sufficient cavity with abruptly sloping
sides, so that water freezing in the vessel will not disrupt
the rock. In the glaciated district in this country boulders
suitable for such use can readily be found, and the cost of
a stone-cutter's labor in excavating a basin to hold six or
eight gallons of water is slight. Where the water is to be
brought in pipes from a considerable distance, care should
be taken to have a tap so placed in relation to the supply
that the pipe will not freeze. The writer's experience indicates that ordinary iron pipe used to lead water from a
spring is likely to rust in a very rapid way. The process
called galvanizing, effected by dipping the pipe in a metallic alloy, prolongs the life of the metal by some years.
A yet better result is obtained by the use of any of the
several enamels which are made to serve as protective
coatings. On the whole, for a stream which is to flow continually to the trough ordinary wooden pipe appears to
be best.

The method of providing watering-places by diversions
of the road on the side of bridges is not to be commended
unless pains are taken to pave the ford with well-matched
blocks of stone. Such watering-places are not serviceable
to heavily laden vehicles, and are, moreover, frequent
sources of accident.

BRIDGES

There are few roads of any considerable length which
do not cross streams of such volume that they have to be
bridged. It is usual in this country to make these bridges
of either wood or iron. Either of these methods of construction entails a large cost in repairs. The best resource,

where the community can afford the additional first cost, is to make the bridges of stone arches. By so doing, though the first expense is the greater, the structure may be a source of no expense even after centuries of use. There are many Roman arches in southern Europe which have withstood the tax of time, and even the efforts of military engineers to destroy them with gunpowder, the strong masonry, owing to the excellent mortar, shooting out the charge as from a gun. If any reckoning is made as to the landscape effect, the value of a stone arched bridge must be accounted as vastly greater than that of any other kind. There are, indeed, no other architectural features attainable in our American landscapes so well calculated to enhance their beauty as the sight of well-shaped masonry arches over the streams. Following any well-chosen model, a country builder can be sure of a success which, however simple, will be of a monumental character.

Where the conditions do not permit the use of stone arches, a less pleasing but very satisfactory bridge, one that is likely to endure for a long period, can often be made by using stone slabs stretched between the abutments and covered with a surfacing of macadam. It is, however, not often possible to make spans of more than eight or ten feet with this system. It is therefore applicable to relatively small streams. Where the bridging has to be done with iron or timber it is generally possible, if the span does not exceed thirty feet, to accomplish the end by means of girders which are placed below the floor. This is particularly the case with iron bridges. If this method is used in structures of wood, the girder timbers, owing to their position, are more likely to decay. Where the bridges are to have a considerable length the state of the art which guides such constructions dictates that they are to be built of steel in a manner which makes them necessarily very unsightly. They cannot be fitly decorated any more than a skeleton. They have to be accepted, if needs be, as detestable inflictions which can only be avoided by the use of the costlier stone arches.

CHAPTER VIII

METHODS OF CONSTRUCTING ROADS

Conditions of roads. Preliminary study of locations. Grades of roads. Surveys and plans. Width of location. Drainage. The hardened way — broken stone. The hardened way — gravel. Cost of maintenance. Macadam roads. Methods of preparing stone. Preparation of the road-bed. The process of compacting. Wear and repairs. Shape of wheels. Annual wearing of roads. Methods of repairing

IN this and the following chapter some account will be given as to the course which needs to be followed in building country roads on various scales of cost, and with the several kinds of material which are accessible in different parts of this country.

It should, in the first place, be noted that the task of planning and building any road in such a manner that the utmost advantage may be had from the conditions of the ground and from the material to be used is one of much difficulty, and demands the services of a well-educated engineer who has devoted much time to highway work. With all other forms of construction it is possible for a man generally well informed in building to apply, in the manner of a copyist, the methods which have proved successful in other places. It is the peculiarity of a road, as compared with other architectural work, that it depends for its utility in greater measure on the topography and other earth conditions than any other constructive undertaking. Therefore the best advice that can be given to those who are about to engage in road-building is that they seek at once the assistance of the most successful highway engineer who can be obtained. Unfortunately, in the present condition of this country, there are few men in this profession whose advice may be safely relied on. It would probably be an overestimate to reckon that there are fifty men in this country who are competent to build good roads, using in their work certain limited classes of materials with which they have become acquainted. Of those of the larger training, who are able to go into a new country and there to plan ways which will be well accommodated to the surface, which will make the most use of the resources which are to be had from the wayside, there are probably not more than a score. On this account it seems desirable to make the statements in the following chapters in such a form that an intelligent road-master may at least escape the principal dangers which are apt to beset him in his work; that he may see how far he should extend his knowledge by experiment or otherwise before setting about his task.

CONDITIONS OF ROADS

In considering the plan for a new road or the betterment of a way now existing, the inquirer will do well at the outset to make up his mind as to the needs which the road has to serve. To do this effectively in a country of rapidly developing possibilities, such as our own, demands a considerable forelooking. The question is not as to the existing traffic over the way, but as to the use which is to be made of it for some generations to come. To make a

8

costly road only to find that the betterment of conditions which the construction brings about leads to the use of very much heavier laden wagons, which in turn require a change not only of foundations and superstructure, but of grades and perhaps of location, is to fail in that duty by the hereafter which is one of the highest obligations of the engineer. Thus in laying out a road which is meant to serve a purely agricultural community, where the volume of traffic is small and the weight on four wheels never exceeds two tons, the grades and bridges may well be contrived in a relatively cheap way. If, however, the effect of the improvement is to develop the use of water-power and manufacturing, where the commercial conditions may soon demand the use of wagons carrying loads of five tons or more, the construction will shortly be found inadequate.

PRELIMINARY STUDY OF LOCATIONS

If the road which is to be built is to be newly laid out, the line should first be carefully studied, if possible on a good map, and in any case by careful surveys which will give the results to be obtained on different lines. Some engineers are afflicted with the notion that they can plan a road by simple inspection. Except it be upon a nearly plane surface, experience shows that this notion is a harmful delusion. A study of the underlying materials should next be made by pits or, better, by cross-sections to show the character of the foundations at all doubtful points. This test is particularly necessary with reference to the wetness of the under earth. Before deciding on the location it should be determined whence the stone for construction and repairs is to be obtained. It is often justifiable to make a considerable detour in order to bring the way near these sources of supply.   In this, as in other matters connected with the way, its future has to be kept clearly in mind.

The experience of the Massachusetts commission shows it is most important carefully to study as to which of two or more existing roads should be taken as the line which it would be most profitable to improve in view of the existing and future conditions of the area in which the improvement is to be made.   Although the plan may be to improve a way already established, a close study of the line will generally show that considerable advantages are to be gained by somewhat varying the position of the way in order to avoid steep grades or bad foundations.   In the first laying out of the roads in a new country needs of economy, or the desire to bring the way near to the house of some influential person, have in almost all cases led to establishments which are now more or less irrational and need to be remedied.

GRADES OF ROADS

The first and most general question which the road engineer has to meet in the greater part of this country concerns the grades of the way for which he is responsible. It is a well-known fact that the loss of energy due to grades increases rapidly with their steepness.   The following table gives a generally accepted statement concerning this loss. It should be said, however, that the capacity of animals to apply their strength diminishes also with the steepness of the grade, so that the loss of efficiency due to the declivity is more considerable than is approximately indicated in the figures in the following table :

TABLE OF GRADE STATISTICS [1] (APPROXIMATE)

Rate of grade in feet per 100 feet in length.	Tendency down the slope in pounds per ton.	Traction power in pounds required to haul one ton up the slope.	Equivalent length of level road for same expenditure of power in miles.	Maximum load in pounds which a horse can haul up the given slope.
0.00	0.00	45.00	1.000	6,270
0.25	5.60	50.60	1.121	5,376
0.50	11.20	56.20	1.242	4,973
0.75	16.80	61.80	1.373	4,490
1.00	22.46	67.40	1.500	4,145
1.25	28.00	73.00	1.622	3,830
1.50	33.60	78.60	1.746	3,584
1.75	39.20	84.26	1.871	3,290
2.00	45.00	90.00	2.000	3,114
3.00	67.20	112.20	2.484	2,486
4.00	89.20	134.20	2.982	2,083
5.00	112.00	157.00	3.444	1,800
6.00	134.40	179.40	3.986	1,568
7.00	156.80	201.80	4.844	1,367
8.00	179.20	224.20	4.982	1,235
9.00	201.60	246.60	5.480	1,125
10.00	224.00	269.00	5.977	1,036

Where the services of a skilled road engineer are not to be had a person fairly well trained in surveying may safely undertake the work, provided he will set about his task in a careful manner. To do this he should first betake himself to some place where good roads are in process of construction, preferably those which are made of broken stone. A week of observation of such work or, better, a share in it as an inspector, repeating the criticisms and keeping the notes of the resident engineer, who may have immediate charge of the construction, will do much to insure a knowledge as to the methods and conditions of the process. If such a field-study were extended over a term of six weeks, and so arranged as to secure a knowledge of highways adapted to varied conditions, such as are now being undertaken

[1] Condensed from a treatise on "Highway Construction," by Arthur T. Byrne, C.E. (John Wylie & Sons, New York, 1892), p. 270.

by the commonwealth of Massachusetts, the observer, if already somewhat familiar with road problems, would be able to make himself fairly ready for his future tasks. Thereafter, with the assistance which will be given him by the better manuals on highway engineering, he may expect, with a share of blundering, to be able to do tolerably good work.

SURVEYS AND PLANS

It cannot be too strongly insisted on that before any work of road-building is actually begun, whether it relate to reconstructing an old way or to making one that is quite new, a careful survey should be made and the lines of the work well staked out. The records of the survey should include a ground-plan of the road, and if it be an old construction one showing the position of the existing traveled portion of it, the sites of buildings, fences, walls, culverts, etc., as well as the limits of the right of way. The proposed new lines to be adopted for the betterment of the location should also be indicated, so that lands to be taken or which are to be abandoned may be seen. A profile of the way, drawn along its center, should next be prepared, which will show in an accurate manner the vertical positions of every part of the road, and on this should be indicated the proposed cuts and fills. Along with the profile there should be prepared cross-sections taken at every station, one hundred feet apart, each showing the original state and the proposed changes of the construction. Where, at points between these normally placed stations, the road exhibits important local peculiarities of form there should be special cross-sections prepared.

With these results of accurate surveys in hand, those

8*

who are to criticize the project, a work which should be done on the ground, should give particular attention to the following described points. The first consideration should be to the grades of the roadway. Within the limits of the prescribed cost, these should, if possible, be brought to not over five per cent., i.e., three feet rise in one hundred feet of length. In places even with main ways the slopes may, to avoid difficult cuts, especially rock excavations, where the material is not of a nature to be serviceable, as broken stone, be left at the rate of six per cent.; but it should be understood that any greater steepness is likely to be prohibitory on the use of the road by heavily laden wagons, and is certain to incur much expense in maintenance, and this for the reason that the destructive effects of the weather and of the traffic increase in a high ratio with the steepening of the slope. A study of this problem of grades on the ground will often show that a slight change in the position of the route will lead to a great amendment of the objectionable slope.

Along with the question as to the declivities of the roads goes that of the equalization of the cuts and fills; this in general can be trusted to any experienced engineer, but almost every project can be bettered by the criticism of any discreet person who will carefully go over the problem in the field. In this revision care should be taken to see that the changes of grade, especially near houses, do not affect the access to them or the drainage in an injurious way. Alterations of even a few inches in grade are likely to afford a basis for troublesome and costly litigation. It is a good general rule to have the results of all such modifications discussed with the owners of the adjacent property, and their claims released, before the construction of the way is begun.

In most instances it will be found desirable to have the surveys made and the plans prepared some months in advance of the time when the construction of the road is to be begun. This will give time for the careful and repeated criticism of the project which true economy demands. If, as is always best, a skilled highway engineer is to be employed, the surveys should be made under his direction.

In many cases it is advisable to have the road graded and used in the unpaved state for one or even two years before the surface is hardened. By thus preparing the foundations a considerable time before the superstructure is built the road has a chance to become firm from the tread of the wheels, an effect which is likely to result in a considerable saving of broken stone if the way is afterward macadamized. This postponement of the work of hardening the road-bed affords also an opportunity to make a more careful study of the grades with experience for a guide.

In European countries, where the roads are controlled by central authorities, there are rules for grades which engineers have to follow. A certain limited allowance is made on the main ways. This is increased as the roads fall into the lower classes of such constructions. In many parts of Europe ascents of several thousand feet are made by roads which have no more rise than three feet in each hundred of length, the gain being made by zigzags or windings of the way. Although there are certain advantages in this system, it is doubtful whether the gain is sufficient to make it worth while for the engineers of this country to adopt any fixed rates of slope for roads. Those who have used the zigzag ways of the mountainous parts of Europe have had occasion to remark that the expense seems in many cases disproportionate to the results obtained. There are important roads in Massachusetts which cannot without an impossible cost be brought to less maximum grades than seven feet in the hundred.

In considering the grades to be adopted reckoning has to be made on the direction of the traffic. In most agricultural districts the greater part of the weight which is carried goes outward to market, the return loads being very light. In such conditions the greater part of the costs should be applied to those slopes which the teams have to encounter with heavy loads. It should furthermore be noted that the maximum grade on a road, if the declivity be of considerable length, determines the transportation efficiency of the teams. It is thus often desirable to expend a good deal of money in reducing a long slope to the least possible grade, even where short pitches have to be temporarily neglected.

WIDTH OF LOCATION

The next question is as to the width of the location. Where it is designed to provide a country road with the simple adornment of plantations or of a forest strip, it is in all cases desirable to obtain a wide right of way. If, however, the margins beyond the gutters are not to be planted with trees or otherwise cared for, the unused portions of the locations are apt to become mere nests of weeds. Furthermore, as before remarked, it is well to consider the probabilities of the road being paralleled or in part occupied by an electrical tramway. If this is likely the location should be made of sufficient width for that need. The proper width under these conditions has been discussed in the preceding chapter. In many parts of this country, particularly in the Western States, the wide, neglected roadsides not only entail considerable loss of arable land, but are a source at once of disfigurement and of damage by affording a nursery for the noxious weeds.

By the readjustment of roads, as has been shown in the experience of the Massachusetts Highway Commission, it is often possible to avoid serious difficulties with bridges. It frequently occurs that some of these costly parts of the road can be dispensed with altogether; in other instances they may be placed in positions where they are safer from freshets. Where new bridges have to be built a careful study of the watershed above them and of the upper limit of rainfall needs to be made. Not the least of the advantages of our system of observations in the weather service is that it permits these reckonings to be made with sufficient accuracy for nearly all parts of this country.

DRAINAGE

After the location of the new or old road has been determined on, the next question which arises is as to the drainage of the way. Whatever material is adopted for the hardening process, the drainage problem remains substantially the same. In general it may be said that the drainage of a road should be planned so that no water should, under any conditions, flow from beyond the road upon its surface, and that the water which falls within the improved way should be discharged into the gutters, and from those gutters to freer channels, as speedily as possible, and, furthermore, that the subjacent earth should be effectively kept dry to a depth varying according to the penetration of frost, but in general for at least three feet below the crown of the road. It cannot be too often said that the surface of a road must be in effect a roof; that the section below it should be kept by that covering in a perfectly dry state, and that it should be protected from the penetration of water sidewise beneath the covering. It is in this portion of the road engineer's work that he finds the greatest need of skill and experience.

There are two ways in which the dryness of a naturally wet road-bed may be brought about. One is by drains upon the side, and the other by a drain in the center. It is rarely possible to make the side drains in the open or gutter form deep enough to insure an effective withdrawal of the water from the central portions of the way. It is generally advisable to place a pipe drain beneath the gutters in the position shown in the diagram. Ordinarily a four-inch earthen pipe is sufficient for the purpose. This should be covered to or near to the surface with a coating of very coarse gravel or, better, pebbles, or, if such material is not to be had, with broken stone such as is used in covering the road. The pipe should, in general, be at a depth of not less than three and a half feet below the crown of the road. (See Fig. 5.)

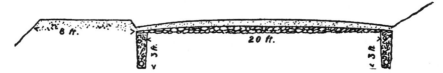

Fig 5.—Macadam road on side-hill showing drains.

Bottom layer consists of large-broken stone, 3 to 4 in. in diameter, 4 in. deep in center.
Top layer consists of broken stone, 6 in. deep in center.

If the road is to traverse a very wet place a better, though in general more costly, drainage may be made by having a V-shaped cavity, as is shown in Fig. 6, below the whole road, placing the pipe or culvert drain, generally over six inches in diameter, in a central position, as is indicated in the figure. Above this should be placed gravel or small-broken stone as before, and the whole of the space

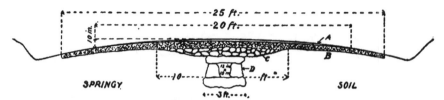

Fig. 6.—Type of road on a very wet foundation.

A=1 inch of stone dust ⅜ in. in diameter.
B=6 inches of broken stone from ½ in. upward.
C=10 inches of field stone.
D=Stones of drain 6 to 8 in. thick.
Total depth from surface 3 feet.
Scale 1 in.=3 feet.

above, up to the level of the hardened way, should be filled in with stone. This stone may be boulders up to six inches or more in diameter, to near the top of the filling. It goes without saying that the drain at the bottom of the structure needs to communicate freely with a permanently open exit. The slope of all such drains as here described should be, if conveniently possible, at least three inches to the hundred feet. To accomplish this result is often a considerable task to the engineer, as the level of the road has to be adjusted in relation to the project.

The open side drains or gutters should have a depth and width proportionate to the amount of water which they have to carry. If the amount of this which comes from the neighboring country on either side of the road be not unusually great, the depression need, in most cases, be no more than thirty inches wide, with a depth of no more than three inches. It is often the best plan to leave the gutters for a few months after the road is constructed, so that the

effect of the water upon their bottoms and sides may be
estimated from experience. Where the conditions are such
as to permit of a grass bottom the channel may be paved
with sod, or often merely "sowed down." Where the

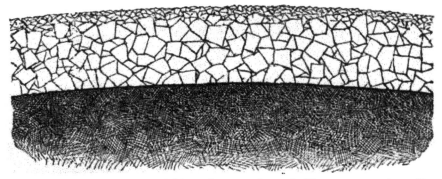

Fig. 7.—Cross-section of Macadam roadway laid on compact earth,
and made solid and permanent by heavy rolling.

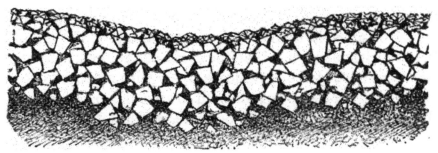

Fig. 8.—Cross-section showing wasteful use of Macadam material.

water cuts the bottom at all it is necessary to protect the
channel with some form of pavement. Where suitable
pebbles abound, those which can be selected having a
diameter of from four to eight inches, they may advantageously be used as a pavement, which should be laid by
a person skilled in the art. Trained pavers can make a
good job with poor materials where dabsters will spoil the
work. Where pebbles for paving are not obtainable some

MATTAPAN STREET MILTON MASSACHUSETTS, SEVEN MILES FROM BOSTON

form of concrete is best suited to the needs. Experience with wooden troughs has been generally unsatisfactory for the reason that the water is apt to find its way beneath the plank, thus leading to damaging washouts. The question of the larger bridges of a road has already been discussed in a previous chapter. The smaller waterways, which are to be provided for by means of culverts, need attention. Great care should be taken to insure sufficient room for easy passage of the floods, so that under no circumstances shall the water be impounded in such a manner that it may flow over the road. The effect of a stream of this sort on a Macadam way is quickly to remove the cementing material and to convert the mass into a rubble, which is apt to wash away, thus leading to costly repairs. The only method to secure this adjustment is a close reckoning as to the drainage area from which the water comes, together with an account of the maximum precipitation.

THE HARDENED WAY—BROKEN STONE

The alinement of grades and drainage of the road having been provided for, the next stage is to consider the form and construction of the hardened way. This portion of the road needs, in all cases where the way is newly made, to be excavated to a depth which will insure the removal of that part of the soil which is much occupied by roots and is of a distinctly open texture. If, as is usually the case, there is a consolidated undersoil or hard-pan a few inches below the surface, this should, if possible, not be broken up, for the reason that it affords a better foundation than can be obtained at a greater depth. Where the construction is to occupy the site of an established way it is very desirable not to break up the firm bed which has been already formed by the pressure of the wheels. If possible, the surface of such a way should be shaped so as to be the foundation for the layers of material which are to serve for hardening. Where convenient it is well to shape a "dirt road" which is to be macadamized some time before the broken stone is placed upon it, so that it may become uniformly compacted by the traffic.

In practically all instances the improvement of a road such as is here considered demands hardening by means of artificially broken stone, or that which is brought to this state by natural processes, as in the case of gravel. It is therefore necessary to consider the preparation of the bed for the application of such materials. The first step in this process consists in shaping the foundation so that it may

Fig 9.—Chestnut Street, Chelsea.

Showing method of macadamizing over old gravel road, without breaking it up. Broken stone, 6 in. deep in center to 4 in. on the sides.

as nearly as possible conform to the outline of the surface of the way when it is completed. With this intention the bed should slope from the center laterally each way, with a grade of about a half-inch to the foot. This arrangement is not to be regarded as imperative. It is, however, advantageous, particularly where the road is to be covered with broken stone. Wherever the nature of the foundation permits it is best to roll the bed with a steam-roller, in order to bring it to an even, compact state, such as can be obtained only by such treatment; by this means not only is the base of the road made the firmer, but the amount of the materials to be used in hardening is more or less reduced by keeping the broken stone from working into the bed, and the under earth is made to aid in up-holding the way.

THE HARDENED WAY—GRAVEL

Where gravel is used as the hardening material it should be chosen with care as to its quality. So great are the differences in the character of this material, and so indistinct the indications of value, that it is hardly possible to give any general directions as to the choice of sources of supply. In addition to the statements already made in the chapter on road-building materials, it may be said that good gravel rarely contains more than one fourth part of sand or clay, that it never shows a trace of slipping after being frozen, and that it rarely has more than one half of its pebbles composed of white quartz. Where the walls of the pit are so far consolidated that they remain steep after exposure for a winter, the material may be assumed as fit even without any other treatment, except that for removing the pebbles which are more than about an inch in diameter. It is rarely the case that gravel can be found so free from pebbles of large size that no treatment to remove them is required. Where this occurs the mass is almost certain to be either so sandy or clayey that it is unfit for use. For the removal of the larger pebbles either of two means may be adopted. The material may be placed at once upon the road, the over-large fragments being then removed by means of rakes, or it may be sifted by ordinary gratings or, better, by rotary screen-drums driven by power. It is much better to have the screening done before the gravel is placed upon the road.

The coarser pebbles which have been separated from the mass of the gravel may advantageously be used for various purposes in road construction; those of fit size for paving gutters, the rest either in covering the drain-pipes, or as a bottom layer of the hardened part of the way. If convenient these pebbles may be used, as for such a layer, say six inches in thickness; on top of these pebbles the screened gravel should be placed to the depth of not less than six inches. If there is no bottom layer of pebbles the total thickness of gravel should be from ten to twelve inches. It is the custom of some constructors to roll the gravel in two successively placed layers; it is, however, doubtful if this method serves any useful purpose. As will be noted in the case of a broken-stone covering, the rolling serves at once to wedge the angular bits together, and to produce by the abrasion of the adjacent bits a certain amount of dust, which acts as a cement to bind the fragments to one another. Neither of these effects is to any great extent brought about by the use of the roller on gravel. The adhesion in that material is effected in most cases by the iron oxides, lime, or powdered stone, with a cementing property which the mass may contain. The efficiency of these agents is not likely to be increased by rolling.

Some highway engineers are accustomed to cover the surface of a gravel road with a layer of clay or loam. Sparingly used, such a covering, say to the depth of an inch, may give to the road a certain temporary firmness which it would not otherwise have; but it is very doubtful whether any permanent benefit is ever thus obtained. The effect of the clay is naturally to prevent or diminish the action of the cementing agents which the gravel contains. Where, as is often the case, a true cementing hard-pan earth can be obtained, it may, if used as a binder, prove serviceable in a permanent way; but as the greater part of such clayey substances fail to bind after being subjected to frost, their use is not to be commended.

When the gravel is placed on the road the surface should be crowned toward the center with a slope of about one half to three quarters of an inch to the foot. As on other kinds of ways, this slope needs to be rather steeper on inclines than on level ground, and this for the reason that the object of the shape is to insure the speedy discharge of the rainfall from the way into the side gutters. On declivities the speedy flow of this water makes it most important that it should be turned from the road. To attain this end on an imperfectly hardened surface, such as is always found on a graveled highway, it may be necessary to face the eminent disadvantages of a steep crown, making the slope as high as an inch to the foot.

As the surface of a graveled road is at best a rather imperfect thing, it is hardly worth while to discuss the form of the crown which should be given to it. In general it serves sufficiently well to continue the slopes from either side to near the middle of the way, without any effort to strike the somewhat elaborate curve which it may be well to give to a road of higher grade of construction. However well built a way of gravel may be, it is sure, after a short period of use, to become rutted by the wheels, as well as channeled by the paths followed by the horses. As soon as this state is brought about the road should at once be repaired. This may be done by filling the depressions with fresh gravel, a process which is expensive, and is likely, especially when frequently repeated, to deform the road by increasing the steepness of the pitch toward the sides. A better mode of repairing is to pass some form of a scraper with converging flanges over the road, so that the materials cast out from the ruts may be drawn

9

back into those depressions. A simple and tolerably effective tool may be made by taking a large, heavy wagon-tire and loading it by a bar or iron or other weight strapped across the middle. This, when dragged by a horse along the rutted portion of the road, will, though rather inadequately, help to mend the way by dragging the material into the ruts. It is much to be desired that the road machines in common use should have an attachment suited to this purpose.

It is characteristic of gravel roads that they take time to come into good condition. Rarely do they show at their best until they have been in use for several years, all the while being subjected to frequent repairs. Given material of good quality, and diligent care in its use, a road of this nature may be made as good as one built of broken stone. Owing, however, to the small size of the fragments which take the blows of wheels and feet, the rate at which the materials pass into dust is, in almost all cases, much more rapid than where the larger fragments of artificially broken stone are used. No good data are in existence to show the relative rates of wear on these two classes of ways, but it appears likely that on an ordinary country road, where the carriages rarely weigh more than a ton and a half, this rate is about one half greater on the graveled way than on that paved with broken stone, and that the expense of occasional repairs is in about the same proportion. On this basis we may safely assume that in money costs it is more economical to build the less used rural ways of gravel than of broken stone. When the sources of supply of these two classes of materials are equally near at hand, the graveled way is likely to cost for the superstructure alone not more than from thirty-five to forty-five per cent. of the broken-stone covering.

The interest chargeable on the surplus of cost of the last-named kind of road would, in most cases, be sufficient to provide for the considerable annual charge due to the larger amount of repairs required on the weaker graveled ways.

COST OF MAINTENANCE

As yet observations as to the relative cost of maintaining these two classes of ways do not permit us to make any definite statements as to the expense which they respectively entail. It is, however, tolerably plain that the greater cheapness of the graveled way makes the consideration of that method of construction fit, wherever suitable material is to be had at a low price, and where the ways do not have to endure heavy travel. It may safely be assumed that, wherever a road has to bear each day the burden of half a dozen or more teams with loads exceeding three tons in weight, the economy of gravel is very doubtful.

As the principal difficulty in maintaining gravel roads arises from the rapidity with which they wear out under the stress of heavy vehicles, it is often well, when the tax of repairs becomes excessive, to cover the surface after it has been brought to the desired shape with a coating of broken stone. This coating, if laid on an old, well-compacted gravel way, need not be more than four or five inches thick after rolling in order to secure the best results; in fact, a little more than one half the stone required to build a Macadam way on ordinary foundations will serve to build the road if it is laid on an established base of this nature. It is thus evident that, even where the ultimate maintenance of the way with this material is doubtful, it may often be judicious to begin the improvement of a country road by constructing it of gravel. If in time the traffic proves too heavy for the relatively weak structure, the stronger top coating may be applied; the total expense of the way in its final state being little if any more than what would have been incurred in case the building had been done with broken stone at the outset of the undertaking. Although this recommendation of experiments with graveled ways may not commend itself to many engineers who look rather to substantial work of the highest order than to the economy which must guide in such matters, the writer is convinced that, with the greater part of our rural ways, the policy as above outlined is worth a careful consideration.

As regards graveled ways it is well to note that, where the cementing capacity of the material is not good, it may often be greatly improved by covering the surface to the depth of about an inch with any of the red or brown iron ores which are so commonly found in nearly all parts of this country. Even where these ores are so impure that they have no value as sources of metallic iron, they may serve excellently well as binding materials to hold the pebbles of gravel in a well-knit mass. One of the commonest and most accessible fields of supply of ferruginous materials is to be found in the layer of limonite ore which is formed in the bottoms of swamps wherever the water which drains into them contains a notable quantity of iron. In by far the greater number of the swamps which lie within the glaciated district of this and other countries this layer is to be found having a thickness of from a few inches to some feet; its presence can be readily ascertained by means of a metallic sounding-rod, such as a crowbar, or, better, by using a section of steel pipe, which may, by driving into the bed, be made to bring up a sample of the

material. In the condition in which they are usually found
these iron ores are quite solid; they need to be broken to
small bits before they are applied to the roadway. In this
shape they will be quickly pulverized by the wheels, and
in part dissolved by the rain-water, so that the material
may enter the road-bed and do the desired work. Where
iron ores are used for cementation they must be, from time
to time, freshly applied. The effect of a coating an inch
or two in thickness may last for four or five years, but the
leaching action may remove the material in less time. From
the observations of the writer, the use of iron ore or cement is especially advisable where the pebbles of a gravel
are mainly of white quartz. Such bits of stone are so hard
and smooth that it is practically impossible to bring them
into a firm-set mass without such a "binder." It is likely
that limestone mingled with quartz pebbles would in a
measure induce the same union of the bits of smooth
stone, but this experiment appears not to be recorded.

At certain points in this country gravels are found
which are already mingled with iron oxide in proportions
well suited to give the desired cementing effect. So far
as observed, such deposits do not occur in any quantity
in the glacial drift; they are characteristically old river
accumulations which have been left above the range of
the streams which laid them down by the downcutting of
the channels. Having been long exposed to the action
of the soil waters, the fragments of iron ore which the
mass contained have been dissolved and disseminated
through it so as to cover all the pebbles. These gravels
are not so firmly bound as to make it difficult to excavate
them. When placed on a road the material quickly com-
pacts into a tolerably firm mass. Although, so far as
known to me, these iron-charged old stream gravels have

9*

been exploited only in the region about the junction of the Ohio and Tennessee rivers in western Kentucky, there is reason to believe that they may be discovered along the banks of many of our Southern rivers. They may properly be sought in any terraces lying above the flood-plains of those streams. Their presence is likely to be indicated on the higher ground by the plow. They are apt to appear in the banks of the streams which cut through the deposit.

Closely akin to gravel are the small, water-worn fragments of rock and the talus accumulations found beneath cliffs or on steep hillsides in the regions south of the glaciated districts. These talus bits, not having been water-worn, retain their sharp angles. They are indeed natural Macadam material of a relatively poor quality for the reason that, while still angular, they are always much softened by decay. Still, in some parts of the southern Appalachians they may properly be sought for a road stone. The only rocks the talus fragments of which retain their shape are the cherts, which at the best are too brittle to endure well, though for a time they make a fairly good but rather dusty way. When used these talus breccias may be treated in the manner followed with other broken stone. It is always best to pass the fragments through a crusher and sizing-drums.

Somewhat allied to the old river gravels are the widespread deposits of pebbly, generally reddish clays which cover so large a part of the Piedmont district of the Southern States of this Union. Although in general these pebbles, from the so-called Lafayette formation, are mixed with clay in so great a measure that they cannot well be employed in road-making without a process of separation, it often happens that considerable masses of the beds are in excellent state for use without any other treatment than the simple process of picking out the bits which are too large to be applied to the road.   It is often the case that the ferruginous matter of these Lafayette beds exists as a coating on the pebbles, and is sufficient in quantity to bring about a tolerably firm binding of the material when it is placed on a road.   In view of the fact that in places where they occur other and better road-building materials are often lacking, these upland gravels of the South are to be recommended for use in the maintenance of ways in the southern part of this country.   They may be used in many places with such economy that a good road, apart from grading, can probably be made for a cost of less than one thousand dollars per mile.

It may be said that the most immediate prospect of bettering our American country roads, apart from the main ways, consists in a systematic utilization of the extensive and varied gravels which are to be found in all the ancient glaciated district and in nearly all of the more southern districts.   Therefore the just aim of intelligent road-masters should be to seek out the deposits of this nature within their respective districts, ascertaining their value and the means whereby the best results may be obtained from their use.   As before stated, little in the way of useful directions can be given that will serve to guide the explorer in this field ; but experiment with the materials in short bits of road is not costly, and if care be taken to use pebbles tolerably free from sand and as free as possible from clay, except it have a very ferruginous character, the results will in almost all cases be in a fair measure satisfactory.   Where the experiment fails it will usually be because the pebbles are of smooth hard quartz, or where, being of other kinds of rock, they are so far decayed that
. they fall to dust when subjected to the strain which they

have to meet in the roadways. It should be understood that the road-master has to make himself in many ways a competent observer of nature. There is no better place for him to begin this work than by the study of gravels, which are, indeed, among the most interesting and varied of all the objects with which the geologist has to deal. In this business he will do well to trust to himself, for, while there are many learned treatises on gravels, there are none which will serve his needs. It is indeed much to be desired that the States which are moving for better highways should have careful studies made concerning the gravels within their several areas, or that the federal Geological Survey should undertake the task for the whole country.

MACADAM ROADS

Where it has been determined to build or rather harden a road by the use of artificial broken stone, the question at once arises as to the source whence the material is to be obtained. As noted in the chapters on the distribution of road materials, there are but few sections of this country where quarries yielding suitable stone exist or can be opened near the site of any road. The varieties of rock which are fit for such use are of relatively rare occurrence. The result is that in most cases the cost of transportation of the broken stone to the way, provided it is obtained by quarrying, is likely to be the largest element in the expense of its application. If the road is to be subjected to a heavy traffic other than that based on agriculture, it is usually worth while to incur this cost at least for the upper layers of the construction, and this even where the charge for transportation amounts to as much as a dollar a ton. But on less taxed ways it is in most instances best to depend .

on cheaper local materials where such can be had of fair quality. For this purpose the road-master should carefully study the sources of local supply; first of all, those which are to be found in the "field stone." By the term "field stone" are indicated the loose fragments of rock which in the greater part of the upland districts are to be found scattered over the surface or accumulated in the river-beds. Except where these boulders are of quartzite of a crystalline nature, with a sugar-like fracture, or of slate, they are, if not too much softened by decay, tolerably well fitted for use in macadamizing.

METHODS OF PREPARING STONE

In the old, generally disused method of preparing stone for road use the breaking was done by use of a light, long-handled hammer, an expert laborer being able to break an amount varying from half a cubic yard of the toughest traps and granites to two cubic yards of limestone per diem. Allowing the wages to be one dollar and a half per day, the expense for a cubic yard would thus vary, according to the nature of the material, from seventy-five cents to three dollars. This primitive method caused the charge for breaking to be in most cases the heaviest element of the cost of broken-stone roads. In the modern system, made possible by the invention of the jaw-crusher, the stone can usually be crushed at a cost not exceeding thirty cents per ton. Contracts for preparing lots of from three to five thousand tons can usually be let at that price. When municipalities own the necessary plant for this work they can usually crush the stone, after it is delivered on the platform, for less than thirty-five cents per ton, allowance being made for interest, depreciation, etc. It may also be noted that there is relatively little difference in the output of a crusher due to the variation in the character of the stone. The range on this account is probably not over fifteen per cent. There is, however, a decided difference in the cost of repairs, which may be doubled or trebled by unusual toughness in the material supplied to the machine. Allowing for these conditions, the cost of crushing by hand, as compared with that by steam, is probably on the average as one to three in favor of the modern method.

It should be noted that there are certain collateral advantages in favor of the hand method which serve, in a way, to balance the account as above rendered. All the existing forms of crushers which have proved serviceable are so constructed that they cannot be conveniently moved. The cost of taking them to pieces, even for a journey of say two miles, including the expense of water-supply, etc., may be reckoned at two hundred dollars. The result is that in practice it is not found profitable to set up the apparatus with the intention of crushing less than about five thousand tons, which commonly requires that from one position the plant is to supply the stone for a mile each way from the site it occupies. This means that the broken stone has to be hauled for an average distance of half a mile to the point where it is to be used. The charge for this hauling may be reckoned at twenty cents per ton. Where the stone is broken on the side of the road for repairs, or in the center of the way when the construction is new, this last-mentioned cost is avoided. Moreover, the power-crusher, especially when operating on field stone, produces a very much larger amount of "fines" and dust than is made in hand-breaking. While a portion of these materials may advantageously be employed in finishing the broken-stone surface of the road, there is often a great excess of the supply, which, except it be used for sidewalks or kept for repairs, is worthless. If retained for later top-dressings of the road it should be housed to prevent decay, a process which entails much expense. Thus the excessive production of "fines," when field stone is used, in the crusher is apt to add from five to ten cents per ton to the cost of the available product, or, in other words, the incidental expenses of power-crushing are likely to bring the cost of the process in certain cases very nearly to that incurred in hand-breaking. It is to be said, however, for the power-crusher that by the system of sizing-drums it brings the product to nearly uniform dimensions, a feature which, as we shall shortly have occasion to note, is one of much importance in the work.

Although the power-crusher has done much to cheapen the cost of Macadam roads, it does not follow that hand-breaking should everywhere be abandoned. In many parts of the country, where the stone is of an easily broken nature, and where at certain times of the year labor is cheap, and especially where it is intended to build but a narrow hardened way of broken field stone, it may be found best to continue the ancient method of the hammer. I am disposed to protest against the assumption that is now so generally made, that no economy in building macadamized roads can be attained without the use of the crusher, mainly for the reason that the great cost of such apparatus will assuredly limit its use to relatively rich communities, while the main object of the friends of good roads should be to promote their construction in the poorer parts of the country. If the erroneous notion becomes established that hand-breaking is

always hopelessly expensive, we shall have a new and unreasonable prejudice to contend against, one which will
serve to block the way to the improvement of our rural
highways. It is a characteristic humor of our time to
worship machinery and to assume that any inhuman form
of power is cheaper and better than that of the hand of
man. In the case of stone-breaking, as in that of many
other arts, the advantage in the way of economy in the
machine over the hand is in many places decisive; but
this is not to be taken as showing that in all instances
the new should be made to displace the old method. It
is the part of the engineer to keep an open mind in all
such matters, allowing the conditions of each problem to
determine the ways in which it should be dealt with.

The most distinct advantage of power-breaking is that
it is possible with that method, as it is not economically
so with hand-work, to size the stone into those dimensions
which can be applied in successive layers upon the road.
Experience shows clearly that the endurance of the way
depends in large measure upon this arrangement of the
materials, for where a fragment say six inches in diameter lies amid bits say only two inches through, it
always tends to work up to the surface. In a less degree
the same action is seen where the diameters of the pieces
differ only as two to one. The reason for this is that in
wet weather, particularly after a time of frost, the stones
of the way are not firmly bound together, but are likely
to rock about as heavily laden vehicles pass over them.
In this movement the larger the fragment the greater its
swing, because of the length of the leverage which it gives
to the stress. In these swayings the smaller bits continually fall beneath the lifted ends of the larger stone, so
that at each movement it rises higher until it appears

on the surface.   In its upward movement it works to dis-
organize the structure of the road.   Where it escapes it
makes a weak place which is apt to become the seat of a
cradle-hole.

PREPARATION OF THE ROAD-BED

The preparation of the bed for the broken stone requires, as in the case of all roads, some careful surveying
to determine and stake out the cross-sections.   After this
is done all the true soil layer should be cleared away; i.e.,
all that shows much vegetable matter, all trees, bushes,
and roots, need to be entirely removed from the bed.   If
this clearing reveals firm gravel, compact sand, or a true
hard-pan such as will not soften when wetted, the road
may be founded on it.   Where the bottom is of clay such
as softens in wet weather it is best to excavate for the
intended width of the hardened road to a depth of about
sixteen inches below the crown-line of the way as it is to
be completed.   In the bottom of the excavation a layer of
gravel four inches thick should be placed, and on this a
layer of rather large bits of stone, set by hand on edge, in
general after the method of Telford.   This layer of stone
should be compacted with the steam-roller.   A cheaper
plan which has been successfully used as an alternative
is to use only gravel in the foundation of the bed, the
layer being made twelve inches thick.   When the ground
is very soft the Telford method is clearly the better, and
in this case the work should be done with large, preferably flake-shaped stones set carefully together, the interspaces being filled by jamming smaller bits into them.
The rough surface of this subpavement should be broken
down to something like uniformity by the use of heavy

hammers, so that no projecting points will come within four inches of the surface of the finished road. When completed this pavement should be arched slightly toward the center, with no more than the pitch which is to be given to the surface, or say about half an inch to the foot.

Fig. 10.—Section of Holyhead Road. Built by Thomas Telford.
Section showing paving foundation.

Fig. 11.—Section of Holyhead Road. Built by Thomas Telford.
Section showing gravel foundation instead of pavement.

Where, as is often the case on the seaboard or in the glaciated districts of the interior of this country, the trench of the roadway is bottomed in soft sand, some hardening of the surface is necessary, else the roller will churn the sand and broken stone together, until the mixture, which has no value whatever in the road structure, has absorbed, it may be, half of the materials reckoned on for the hardened way. To hold the stone and the sand apart it has been the usual practice to cover the sand with a layer of pebbles of conveniently large size before the bottom layer of broken stone was laid down. This is an expedient which is often costly and sometimes impracticable from lack of fit materials. An experimental inquiry into the conditions of the movement of sand under pressure led to the conclusion that an arrangement which would prevent the sand from mingling with the stones, for the brief time required for the passage of the roller in its first traverses over the road, would attain the desired end. It is not at all needful that the partition should be enduring, for as soon as the lower layer of stones has been forced into contact, and has become bound together, there is no further danger of the mingling of the bits with the sand; thus the speedy decay of the fabric is a matter of no consequence. As the results of careful tests made by Mr. Charles Mills, the chief engineer of the Massachusetts Highway Commission, and with the assistance of Mr. W. P. McClintock, the engineer member of the board, it appeared that ordinary cotton cloth of the cheapest quality, such as goes under the name of cheese-cloth, if spread upon the sand after the road is shaped to receive the broken stone, will serve to keep the stone and sand from churning together. This method was carefully tried in macadamizing the State road between Cottage City and Edgartown, Massachusetts. The cloth was spread in strips lengthwise of the way; the stone for the bottom layer was shoveled from the sides upon it with no unusual care. When the roller came to be used it was found that the stone acted essentially as if it was on an ordinarily firm foundation; it at once united with the usual number of passages of the roller over it. At the present price of cotton, cheese-cloth can be had in large quantities at a cost of about three cents per square yard on the road. This for a hardened way fifteen feet in width amounts to about seven hundred and fifty dollars per mile, which is often much less than the cost of any other effective means of attaining the object, and may be less than one third that due to the loss of the broken stone which would occur if it were allowed to come directly in contact with the sand. A section through such a "petticoat road," as it has been termed, shows that the stones do not tear through the cloth. It is indeed probable that material of even slighter texture and of much less cost would serve. Various kinds of strong paper were tried, but found worthless.

THE PROCESS OF COMPACTING

When the foundation for the stone has been prepared the first layer of fragments may be spread. Where trap or other rock of like hardness is used the lower stratum may be made six inches in depth before rolling, the fragments by screening at the crusher being brought to a range in diameter from one and a quarter to two and a half inches. After rolling this layer will be diminished in thickness to about four inches. It is best to have the length of the strip thus metaled not less than two hundred feet. The spreading should be evenly done. When the broken stone is dumped from carts it should fall on wooden platforms; otherwise the part which remains after the rest of the heap is spread, being somewhat more compacted than the rest, is likely to make a higher place in the finished road. The roller should now be passed over the surface, beginning on the edges and working toward the center. This should be continued until the mass is firmly set, until it ceases to move under the tread, until, indeed, a stone of the larger size used in the layer will be crushed by the instrument without indenting the mass. Still in this state it must not be expected that the stones will be so firmly set that they will not shake under the tread of a man.

On top of such a trap or other hard foundation-stone, which by rolling will have come down from its original thickness in the loose state of six inches to about four

inches in depth, the second layer of three inches thick of
stone, the fragments ranging in size from one half to
one and a quarter inches, is to be placed. This is to be
rolled as before until it exhibits the same evidence of
due firmness. As with the first layer, all irregularities of
surface which may be developed by rolling are to be corrected by filling in the depressions with stone of the same
size as that in the body of the layer. When this second
layer appears firm and smooth, a coating of "fines," or
fragments from the crusher up to half an inch in diameter, is to be spread to the depth of about half an
inch. The roller is then to be passed over this last layer,
with the result that the bits will be ground to powder.
At this stage the road is to be sprinkled with a watering-
cart, but one with fine apertures in the pipes, the work
being done in several passages. The roller is then again
to traverse the way until in its movement the water is
forced upward or pushes before the drums of the machine.
The aim of this combined rolling and watering, after a
coating of dust has been applied, is to insure the entrance
into the cavities or voids between the broken stones of a
sufficient amount of the powdered rock to act as a cement,
which serves to hold the mass together. In part this dust
is made where it is most needed, at the contact of the bits
with one another as they are ground together by the roller;
but experience shows that this amount is insufficient to
give an adequate binding effect.

It might naturally be supposed that this introduction
of cement should be carried so far as to fill in all the
voids between the bits of stone down to the foundation.
Practical tests, however, show that this is unnecessary
and probably undesirable. If the upper three inches of
the macadam are well bound together, it will suffice for all

10

the needs of strength. The open condition of the lower part of the mass promotes drainage and keeps the hardened way in a somewhat elastic state. As the upper layer wears away the dust will wash down into the lower previously open spaces in such a manner as will keep the three or four inches just below the surface in a well-cemented state.

Fig. 12.—A General Cross-section of Macadam Streets.

AA = Gravel sidewalks.　BB = Loam embankments.

Where the broken stone is of a rather soft nature, as is usually the case where the material has been gathered from the fields or stream-beds, it is perhaps desirable to place it on the road in three different layers, arranged as before in the order of the sizes of the fragments, the first two layers being each about four and a half inches thick before rolling, in each case the bits being from half an inch to two and a half inches in diameter, the lowest layer being rolled, as before, before the second is placed upon it. The surface of the upper layer after due rolling is covered with screenings, which are rolled and watered as already described in the account of the treatment of a way built of the harder kinds of stone. Although this system has the advantage that it provides larger and therefore stronger fragments of stone to take the pressure of the wheels, it is open to the objection, previously noted, that these larger bits tend to work upward to the surface. It is questionable whether the advantage due to the evident gain in resistance to the crushing action of the wheels is not more than offset by the risk of the broken stone becoming loosened.

When a broken-stone road has been brought to a normal condition of surface the observer, by sweeping away the coating of dust, will see that the surface has the aspect of a rude mosaic, the flat faces of the bits being crowded against one another so that the interspaces which are filled with cement occupy more than about one third of the area. As the wheels pass over this mosaic the horizontal surfaces of the fragments take the impact and uphold the burden, while the softer cement yields and crowds to force its way downward. The abrasion of the stone which takes place under the wearing action of the traffic alone on a well-shaped road is, where the stone is fairly hard and well kept in position, by no means rapid. It is, however, much affected by a variety of other conditions: those of climate, form of wheels, etc. These conditions we have now to consider from the point of view of the maintenance of the way.

WEAR AND REPAIRS

It is characteristic of roads, as compared with other art products, that it is relatively very difficult to find out the share which the various depreciating agents have in degrading their conditions. It is easy to tell to what the instability of a building or the excessive wearing of a steam-engine is due. Not so, however, in the case of a road. Simple as the thing looks, it is, as regards the influences which affect it, perhaps the most complicated and locally varied result of man's labor. In analyzing the causes of wear on a Macadam road we find that we have in general to assign the first place to atmospheric agencies.

These operate in several ways. Where the bed freezes and thaws the effect is to break up the union between the fragments and to expose them to the decay arising from the penetration of water into the layers. If the mass can be kept fairly water-proof the effect of this action in decomposing the road material is practically limited to the upper surface of the stone. It is to be noted that the mud formed on a much-traveled road contains much dung, so that the water which enters the mass has, because of various acids which are derived from the organic matter, a considerable capacity for dissolving the rock. It is on this account that it is desirable to keep the way reasonably free from mud. In its dry state the powdered mineral does not harm the road on which it rests, though if allowed to accumulate to a considerable depth the coating adds to the resistance encountered by the vehicles; when wet it is an agent of damage which needs to be considered.

While the chemical action of water in bringing about the decay of the stone is important, its effect in washing away the protecting coating of dust is also damaging. On grades of considerable declivity this effect is to deprive the stone of the continuous supply of cement which is required to keep it in place; hence the "raveling out" which so often occurs on steep slopes or on roads which are too abruptly crowned. We thus see that it is desirable to reduce the inclination of ways, not only for the effect on traction, but for their better preservation as well. We see thus how important it is to keep the whole structure of the road from excessively steep inclines. The reader must not suppose, however, that a perfectly dry way would be desirable. In cases where Macadam roads are carried through tunnels we may note that the stone does not hold well together.   There is need of occasional wetting, which may enable the "fines" to work downward and come into a state where they may, when dried, act as a cement.   This consideration obviously throws some light on the matter of watering a road to keep down the dust.   While this is desirable in periods of continued drought, where the road shows signs of going to pieces, the constant use of the method tends to increase the rate of decay of the stone by softening it, so that it readily gives way under the blows of the wheels and hoofs.

In European countries, where the climate is prevailingly humid, the evident damage from mud has led to the custom of scraping this dust away as fast as it accumulates.   Such care is much less necessary in our drier climate, where the winds are apt to remove the excess of the powdered stone which is not washed away by the torrential rains.   Still, on roads having a heavy traffic, when the wearing of the stone amounts to as much as an inch a year this precaution may well be adopted.   When the practice of scraping or sweeping the road is followed care should be taken to secure the removal of the dust after it is heaped on the shoulders of the road; otherwise it will quickly become distributed again over the surface.   In many cases this material is of sufficient value as manure to induce farmers to remove it at their own cost.   Where this is not the case it should be so placed that it cannot return to the paved way or to the gutters and drains of the road; otherwise it is likely to do more injury than if it were left to be disposed of by the winds and rains.

The wearing of Macadam roads in this country due to violent rains is a distinct and somewhat peculiar evil.   It leads often to the removal of the binding dust between the top stones in such a measure that they become loosened.

10*

This points to the precaution of having the slopes so arranged that the water may be carried to the gutters so directly that it will not have a chance to gain the cutting energy which it will possess when it has a speedy flow. On this account it appears necessary to have the side pitch of the surface rather steeper than is prescribed in European works on road-making.

The action of strong winds, in connection with long droughts, which is a common feature in all parts of this continent, often leads to the excessive removal of the dust. Thus in Kentucky, where the southwest counter-trades often blow with great energy and continuance, the limestone roads are often swept clean for weeks of all dust, so that the occasional rains do not provide for the renewal of the cement in the crevices between the bits of stone. Watering in the measure required to prevent this injury would, in such cases, be advantageous. This process is, however, costly. Properly carried out, the annual expense of such treatment cannot usually be brought below seventy-five dollars per mile. The reckoning has to include, on most country roads, a provision in the way of windmill pumps with large tanks to contain a sufficient store of water for periods of calm. It is likely to be many years before this refinement in the care of roads is generally adopted in this country.

SHAPE OF WHEELS

Although the injury done by the traffic on ordinary rural ways is probably less than that effected by the weather, it is of a more evitable nature. Much of it can be avoided by a proper care as to the form of the wheels and horseshoes, the loading of vehicles, and the path they

hollow in the road.   These points we will consider in
succession.

First, as to the width of the wheels.   Where these
are small, i.e., of less diameter than, say, thirty inches,
their effect is to push bits of loose stone before them,
particularly when the vehicle is heavily laden, in such a
manner that the fragments plow up the road until they
have worn out or have glanced aside.   A wheel thirty
inches to three feet in diameter, or less, will do this when
one four feet across will ride over the obstructing bit,
crushing it or driving it down into the bed.   If it were
possible (it is obviously not so) to have wheels limited in
diameter to four feet, good roads would be more easily
maintained in order.   Fortunately there is a tendency
toward the general adoption of large-wheeled vehicles
wherever the roads are made good.   There are, indeed,
several mechanical reasons why this should be the case.
Moreover, in this country, where the carriages are as much
better as the roads are worse than those of the Old World,
it is the custom to avoid the road-destroying, small fore
wheels of farm-wagons which are so commonly used in
Europe.

While the diameter of wheels has been but little considered, the matter of width of tires has been made a subject of much remark.   There has, indeed, been no end of
idle talk concerning this matter, much of it directed to the
point that our American wagon-builders have shown a
lack of judgment in building with narrow tires, while they
should provide their vehicles with broad treads such as are
in use in Europe.   The fact is that in this, as in many
other ways in which our people have departed from ancient
and old-world customs, they have been led by wisdom and
not by folly.   This will on a little consideration be made

evident. Where, as in ninety-nine hundredths of the mileage of American roads, there is no definite pavement the wheels have in muddy weather to descend into the earth until they find a firm foundation on which to rest. In so doing they have to cleave sticky mud which often has a depth of a foot or more. If these wheels were broad-tired the spokes would also have to be thick and the fellies wide, so that the aggregate holding power of the mud upon the vehicle would be perhaps twice what it is at present. It is useless to talk about the advantages of a broader tread to the wheels of our wagons until we have a thoroughly good system of roads which they are intended to traverse. Any laws looking to this end would be disobeyed because of private needs so general that they would amount to a public necessity. When the roads of a district are made good only as to the main lines of communication, the side roads still demand the peculiar advantages afforded by the narrow tread. It is thus only when the good ways are developed to a complete system that the people can be justly required, or even expected, to adopt the proposed broad tires.

While it is clearly injurious to a road paved with broken stone to have very heavy-burdened wagons with narrow tires pass over it, there is no reason to anticipate that such vehicles will continue to be used when the general conditions of wagoning are such as will make it practicable to use wheels of broader tread, except, perhaps, in cities, where there may be some advantage arising from narrow-rimmed wheels for the reason that they fit into the tramway tracks. The greater strength of the wide-framed wheel, with the resulting broad tire, is certain to commend it to general favor. We have here a natural influence which is likely to prove far more effective than any statute.

The best argument against the enactment of laws concerning broad tires is found in the fact that the numerous and long-enforced English statutes on this matter have of late years been abrogated, a century of experience having shown that they were difficult to administer and generally disadvantageous.   In this country, where the machinery for administering such laws does not exist,—where, indeed, a fit mechanism for their enforcement cannot well be contrived without serious changes in our police system,—it does not seem desirable to enter on legislative endeavors which the mother-country has abandoned.   The only fit resource appears to be to trust to the construction of roads in so solid a manner that they will not give way under such strain as unreasonably narrow tires may bring upon them.   In fact, the additional tax on the endurance of a well-constructed road which is likely to be imposed by the inadequately small bearing surface provided by the wheels is probably small. If the road be kept smooth so that the load applies only a steadfast pressure and does not deliver blows to the stone, all but that of very soft nature will meet the strain without damage. When the wheels break through it in almost all cases indicates that the bed has been allowed to wear to a dangerous state of thinness, or that the stone was not put on under such conditions as would insure a proper bond to unite the fragments of which it is composed.   In many cases the damage is due to the fact that care has not been taken to secure a proper foundation where the road is laid upon clay.   As before noted, where clay beneath a road is not kept perfectly dry the structure is always in peril.

There is another kind of violence which wheels do to a road that is brought about by the need of braking or, what is worse, locking them as a laden vehicle goes down a steep grade. Either of these means of hindering the descent of the vehicle inflicts damage in proportion to the extent to which the free turning of the wheel is resisted and its movement converted into sliding. If the bond of the stones is exceptionally firm the action may be limited to a very rapid surface wearing, probably some score times as great as where the wheel is free. If, however, the union of the stones is weak, as it always is either in wet weather

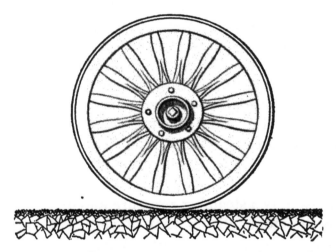

Fig. 13.—Showing wagon-wheel sustained at single point of contact on hard, smooth surface of compact Macadam or Telford road.

or in very long droughts, the effect is to tear out the bits. As in the matter of narrow tires, it seems useless to seek to remedy this evil by laws, and this particularly for the reason that where the load is held back by the pressure of the horses on the breeching the calks of their shoes are likely to do about as much damage as the checked or locked wheels. If the grades are too steep the road must take damage from the defect of the plan; it matters little in which way it is inflicted. The remedy is to diminish the slope to the point where the road will not tend to acquire in its downward course a velocity which needs to be checked by any considerable exercise of power; for all the energy thus used is directly applied to the destruction of the costliest part of the way.

The damage done by the pressure of the wheels on a well-built Macadam road is usually much less than is inflicted by the shod feet of horses, especially where the shoes

Fig. 14.—Showing wagon-wheel pressed into the surface of a dirt road, enlarging the area of contact and impeding the forward movement of the vehicle.

are provided with the usual calks at toes and heels. The stress applied by the tires is wholly downward; when the road has the normal covering of dust this acts as a cushion to distribute the pressure. If the wheel be of considerable diameter, say four feet or more, the area of the work may have a length of two or three inches at any moment of the action. Though the time during which this pressure is accumulated seems short, being, in fact, only a small fraction of a second, it is long enough to permit the elastic mass to be slightly compressed, so that it fits in a measure to the curve of the tire and so sustains the weight without rupturing the cement which unites the fragments of stone. It is quite otherwise with the blows which the feet of draught animals apply. These strokes are delivered suddenly, so that the bed has no time to yield. If they were struck with the naked hoof the elasticity of that part would provide the needed spring, but the heavy shoe is like a stone hammer at the end of a flexible handle; it is a breaking tool of much efficiency. Worst of all, the calks act like wedges to force apart the stones and lift them to the surface. This result is greatest where the grades are steep. In ascending or descending the slopes the calks are driven in and strained in drawing forward or holding back the loads; hence a shearing movement which tends to pull out the fragments of stone.

An inspection of a clean-swept Macadam road which has been some time in use will show the dents which are due to the horses' feet and the fractures of the stones effected by those agents of wearing. Where these effects are distributed evenly over the pavement it may be observed that the wheels tend to heal them by pressing the materials back into place; but in the center of the road there is a space of some width where the wheels rarely do their work and where the hoofs are most effective. Here the action of the horseshoes is often exceedingly injurious; the stone is picked up and beaten to dust, which blows away, so that in a short time there may be a deep foot-rut down the center of the road. To avoid and mend this kind of damage taxes the skill of the road-master more than does any other part of his craft.

The entire prevention of ruts is impossible. They are, however, to be in part guarded against by careful rolling of the center of the roadway. Owing to the necessarily convex form of this portion of the construction, the roller does not bear as evenly on the surface as is to be desired. Hence without much care the center of the way is likely to be imperfectly compacted. Attention should be given to the work in order that the passages of the machine may be numerous enough to bring the materials to a firm state. As soon as ruts begin to appear they should be repaired. Those formed by the wheels may in most cases be healed by filling them with small stone of the size used just before the dust layer was applied. This material will be brought down by the pressure of the tires. The foot-ruts are less easily mended. With them the best plan, where the depression is still slight, is to fill them in with the size of broken stone applied in the uppermost layer, using the roller to compress the mass as in the original work of building the way. Where this use of the roller is not conveniently possible the partial obstruction of the center caused by the broken stone which has been applied will usually cause the driver, or the horses of their own instance, to take to the side of the road, so that the wheels serve to tread the material into place. Where the foot-rut has been neglected and has been allowed to become deep it may be necessary to remove the rounded stone and dust, refilling the space with fresh angular bits. The path then needs, for its future good service, to be carefully rolled.

It is the plan of some road-masters to provide against central wheel- and hoof-ruts by placing, from time to time, obstructions in the center of the road so as to force travel on to the sides of the way. Although this is an effective resource, it is not to be commended on a way of ordinary width; for the harm that the vehicles do in passing one another, being thus crowded upon the shoulders of the way, soon brings that relatively weak part into a state requiring repair. Moreover, these barriers are sources of danger to those who journey by night. The essential feature of a highway is to be at all times freely and safely passable. It is therefore not fit to limit its utility and safety in any avoidable manner.

ANNUAL WEARING OF ROADS

Besides the special wearing of the wheel- and hoof-ruts the general surface of the hardened roadway is subjected to an annual reduction of thickness throughout its whole extent. The rate of this wear is a result of a complicated series of interactions in which the travel (in terms of hoofs, wheels, and total weight), the nature of the stone, and the climate all enter. The computations of Morin, Codrington, and others, based on carefully ascertained data, have given interesting results, which, however, are hardly applicable to American conditions, except in a very general manner. The last-named writer has shown in tabular form the results obtained in some important tests made by M. Graeff ("Annales des Ponts et Chaussées," vol. ix., 1865), which seem to prove that the rate of wearing increases more rapidly than the increase in traffic; the annual consumption of stone for diverse weights being as follows:[1]

[1] Thomas Codrington, "The Maintenance of Macadamized Roads," 2d ed. (London, 1892). See this admirable work for many other matters concerning the endurance of roads.

Daily traffic in tons.	Annual consumption in cubic yards of materials per mile.	Annual consumption per mile per 100 tons' traffic per diem.
1,378	724	52
1,772	1,857	104
2,264	2,780	122
3,150	4,615	146
5,315	9,886	186

It appears from this table that more than three times as much wear per unit of weight took place with the heavier burden of traffic as with the lighter.   This is probably to be explained by the fact that with the greater stress there was no chance for recementation to become effective.   As these tests were made in a road built of broken schist, they probably give a larger factor for the depreciation than would be found true in the case of ways built of less friable and better cementing stone ; yet there is reason to believe that in a general way, but in a less proportion, the increase of traffic in practically all instances heightens the wearing per ton of weight passing over the road.

In the case of the road observed by M. Graeff, as above noted, the wear under the maximum traffic was at the rate of two feet of thickness of compacted road material per annum.   This may be taken as a maximum.   The minimum on a well-built rural way formed of trap, where the daily traffic averages one hundred "collars," [1] may be reckoned at from one fourth to one half inch per year.   Of this loss a portion, varying in quantity with the character of the stone, will be due to the decay of the rock.   If the travel on the road is so light that less than half of an inch of dust is formed each year, there is a chance that the supply of dust will be insufficient to provide for the renewal of the cementing material which is lost by the action of the rain and wind in removing the dust from between the top stones, so that the road may "go to pieces."   On this account a way which has only slight use, as on private grounds or in neighborhoods where there is only summer

[1] The term "collar" is used by European writers to designate the single draught animal.   It is a convenient addition to the already rich argot of the road-master.   It may well be adopted in this country.

travel, in most instances should, if possible, be made of other material than broken stone.

METHODS OF REPAIRING

If a road could be induced to wear evenly downward the task of repairing would then be limited to occasional extensive reconstruction work; but, whatever be the rate of wear, the inevitable irregularities in the strength of the various parts of the construction will lead to the formation of cradle-holes of varying areas. The aim of the road-master should be to detect these as soon as they begin to form, so that the destructive process may be arrested in the beginning before the cavity has been enlarged by the blows given by the wheels as they fall into the depression, a process which goes on with accelerated rapidity with the increase in the depth of the depression. Where the pitting is slight, say not over an inch in depth, and of small area, say a few inches across, the mending can often be done, especially where the macadam is of a rather soft nature, by applying an amount of finely broken stone,—the bits being about half an inch in diameter,—preferably when the road is wet, the coating being laid on so that it is, in the loose state, rather higher than the adjacent surface. The mass can then be driven down with a hand-rammer such as pavers use. Care should be taken to clear the mud from the depression before the stone is placed in it.

A better plan of repairing, one which should be followed where the depressions are already wide and deep, is first to clear away the mud or dust; next to pick up the stone to the depth of two or three inches, placing the new material on top of the loosened mass. The surface of the mass should be rounded upward so as to remove the crown of the pit, with a reckoning that it will shrink about twenty-five per cent. in bulk when compressed. This may be driven down with the rammer, or the compacting may be intrusted to the tread of the wheels; but it is better to have the work done with the steam-roller, so that the top coating of "fines" may be worked in and the place be brought into the same condition as the other parts of the way. A failure to attain this end is likely to lead to a speedy return of the trouble at the same point, so that a permanent sore spot may be formed in the way. If the area thus repaired is small it is not necessary to apply the top coating of "fines," as this material will be brought in from the neighboring parts of the road.

For all the directions which may be given in the matter of road repairs, the result depends mainly on the skill of the road-master in contriving the work so that the surface of the mended place may be in the end exactly in its proper level. If higher or lower the effect is only temporary. A pit will form in the old position if it be too low; other pits will form alongside of it if the place be left too high. The ability to do such work effectively depends upon natural craft as well as training in the art.

The process of patching, at first limited to small bits of the road, beginning, it may be, some three or four years after a good piece of construction has been turned over for use, gradually involves larger areas until in the course of time, say ten or fifteen years on a road of average use, the whole of the way has been gone over at least once, some parts of it several times. At this stage in the life of the way the time has come for "general repairs," when the pavement has to be made over again in a manner which will use the remains of the paving material in the

11

new construction. Before setting about the remaking of the hardened way the need of changes in materials, grades, water-passages, underdrains, width of pavement, etc., should be carefully considered, as also, if they have to be dealt with, such questions as location of water- and gas-pipes, street-railways, and other obstructions. As the road is to be broken up to its foundations, the task should be preceded by a careful revision of the situation, so that provision shall be made against change of plan or any remaking of the way for a long period in the future.

If a proper record has been kept as to the history of the imperfections of the road it will be found that weak places due to bad under or surface drainage or to excessive grades can now be remedied. In anticipation of this need, it may be recommended that for any road there should be a plan on the scale of at least one five-thousandth, so that the position of defective places may be noted thereon for service in the time of general repairs. This record should be revised just before the extended reconstruction is begun.

In the process of reconstruction the first step is entirely to disrupt the pavement to the bottom of the lower layer of broken stone. This may be done by means of hand-picks; it is sometimes accomplished by the use of plows, but this method is objectionable for the reason that it disturbs the foundations and risks mingling the under materials with the broken stone in a measure unfitting the latter for reuse. The best method of accomplishing the work is by means of the steam-roller. The drums or wheels of this machine are provided with holes into which spikes may be fitted. When these are in place the effect of frequent passages over the road is to separate the bits from one another, restoring them to the state in which they were before the mass was originally compacted; in this condition they can easily be separated by hand-picks. When this process is effected with teeth of proper length in the drums, and with a road which has been maintained of tolerably even thickness, there will be no considerable mixing of the foundation material with the broken stone. If this basis is of gravel, as it should be preserved, what may be mixed with the stone will not prove injurious. If there be Telford pavement or cobble foundation it should not be broken up.

After the layer of broken stone has been picked up it should be made even with the same care as when first laid down. The slope toward the sides should be restored; the mass should be brought to the thickness designated in the case of original construction, if necessary by the addition of more broken stone. The mass is then to be rolled in the usual manner. The subsequent treatment of the road is to be conducted as advised for the process of newly building a way.

CHAPTER IX

METHODS OF ADMINISTRATION OF ROADS

French method of supervising roads. English method. Other
European methods. Best system of administration for American roads. Commonwealth system. Methods of inspection.
Removal of snow. Need of elaborating methods of supervision
fitted to the conditions of this country

IT is easy to prescribe the process by which a road is to
be kept in good repair, but in practice it has been in
all countries found difficult to insure the proper systematic
care in the administration of such work. In this country,
where the social conditions do not lend themselves to the
effective execution of any government regulations, the
task presents greater difficulties than elsewhere. Let us
see what instruction can be derived from the methods followed in the Old World, where alone has the highway
system been brought to a satisfactory state.

FRENCH METHOD OF SUPERVISING ROADS

In France, where there is the most extensive and successful system of stone roads, all those of importance are
under the control of the Bureau of Bridges and Roads.

176

This bureau employs all the men, known as *cantonniers*, who have to do with the immediate work of repairing the ways. To each of these men is allotted a section of a way which it is supposed that with his individual labor he may be able to watch over and repair. He is expected to break and apply the stone needed in repairs, to attend to the ditches and drains, to dress the sides of the road so that they may be always in an orderly state. He provides his own tools and simple uniform. Men of this, the lowest, grade are required to be on the road, from May 1 to September 1, from 5 A. M. to 7 P. M.; for the rest of the year from sunrise to sunset. They are required to take their meals on the road, and for this they are allowed two hours each day, except in very hot weather, when they may have three hours. The average working time is about ten hours. For this service the men are paid about fifty cents per day. They are required to be about their work in all kinds of weather, but are allowed to use shelters, which they make for themselves. The cantonniers furnish their own tools.

Next above the simple laborer in this corps comes the chief cantonnier, who has charge of at least six subordinates. This chief also has charge of a section of road, but one relatively short, in order that he may have time to inspect the work of his subordinates and advise them as to their duties. Above these petty chiefs come the officers of the great corps of the "Ponts et Chaussées." In addition to the sketch of the cantonnier system as above noted, it may be said that the force is largely composed of discharged soldiers who have had a disciplinary training. They are fined for any neglect of duty, and may receive annual gratuities amounting to as much as a month's pay each year for particularly good service. As a whole the

11*

efficiency of the corps is excellent. The men work in a plodding, pottering way. The product of their labor is probably not more than two thirds of what would be obtained by a skilful private contractor for the same hours, but it is about as good as would be secured in public work in this country. It is interesting to note that, notwithstanding the low price of this labor, the annual cost of maintaining the main or departmental roads of France

A French national road.

was in 1860 about one hundred and thirty-five dollars per mile. The care of these ways was, however, of the highest order, such as we cannot hope to attain in this country for some decades to come. It may be said without danger of contradiction that good as the French system clearly is, it cannot be applied to a decentralized government such as our own.

ENGLISH METHOD

The existing English method by its system of local administration by counties, through their county councils, represents the result of many experiments in the management of roads other than those in cities. These counties are, as regards wealth and population, fairly comparable with our own States.   In area and concentration of taxable property they are not so, the resources per square mile being many times what they are in any but a few of our communities.   The plan of the English system has been found in general so successful that it has been extended to Scotland and Ireland.   The important feature is the mode in which considerable groups of roads are committed to the care of local surveyors, who are endowed with much authority and who report directly to the county board. Under these surveyors there are foremen and ordinary laborers essentially on the plan of the French cantonniers. As in most other matters of local government in Great Britain, there is a considerable diversity in the details of the road administration in different parts of the country, but the plan appears everywhere to contemplate lodging the authority with the surveyor.

According to Codrington (op. cit., p. 154), a surveyor who keeps a horse and who has foremen under him may be expected, provided his field of duty is compact so that his task does not take him far away from home, to superintend, as they do in South Wales, from eighty-six to one hundred and forty-five miles of way.   The impression left by the statements of the English experts is that where more than about one hundred miles of road are under the care of a single inspector, even if he has the help of good foremen, the money loss from insufficient intelligence in the direction of the work is considerable. It appears to be assumed that the ordinary pay of a surveyor is to be about five hundred dollars per annum, provided he does not keep a horse, and seven hundred and fifty dollars if he provides such means of transportation. At these wages it is estimated that the cost of the supervision alone will be in general from seven to eight dollars per annum for each mile of road. As the pay of men of the surveyor class would in this country have to be about double that which they receive in England, we may reckon the cost per mile as nearly twice that above given.

OTHER EUROPEAN METHODS

The systems of control in Switzerland and Germany are in general much like those of France. They are evidently, from their centralized character, not fitted to guide the people of this country in devising a plan for the control of their highways. They all, however, exhibit certain common features which are evidently the result of long experience in road management, and which it is well to bear in mind. They more or less adequately provide for the permanent employment of laborers who are to have charge of sections of the way through the year. In some states, particularly in France, these men are allowed to be away from their work for a short time in the harvest season. This is done in order that these employees may be able to help in the critical period of farming. All these plans contemplate the employment of foremen and the supervision of the gangs by competent surveyors, who devote their time to this kind of work. The differences relate to the control of the men of the surveyor's grade, which in most countries is provided for by an elaborate administrative system, which is managed by the national government. In this regard the British system is evidently the one that may best be copied in this country. It provides for the actual management of the ways without involving the centralization of authority which is repugnant to all people of the English stock.

BEST METHOD OF ADMINISTRATION FOR AMERICAN ROADS

Judging by the experience of other countries, and such little as has been had in our own, the best system of road control for our needs and conditions is one organized on the unit of the State government. A lesser unit, such as is afforded by the counties, is inadvisable for the reason that it is too small for effective administration. In the several States there should be a board having full authority in all that relates to those ways which are to be regarded as of general importance. It seems best that this body should be a general Board of Public Works, and as such charged with all the engineering business of the commonwealth, and this for the reason that there is an intimate relation between the public roads and all the other constructive undertakings of a State. There is, however, at present a strong tendency to morselize the administration of our State governments, placing each definite task in the hands of commissions. Assuming that there is to be a highway commission in charge of the road business of each State, it should at least have its field of duty include bridges and street-railways, telegraph and telephone lines, as well as the ordinary ways. It should have the right to determine where street-railways should be allowed a place on roads, and also where and under what conditions pipes should be placed therein. This extension of authority is clearly necessary in order that the main object of the board may be attained.

COMMONWEALTH SYSTEM

At present the roads in the several States are, with rare exceptions, held under local or corporate control by the counties, towns, or turnpike companies. The method by which these ways are to pass into the hands of commissioners representing the commonwealth should be carefully provided for in the laws organizing such boards. As far as the public or non-toll roads are concerned, the best way of passing them into the hands of the commissioners seems to be that adopted in Massachusetts, where the motion originates in the local authorities. To have the State board determine the matter altogether at its own instance would, on many accounts, be undesirable. As regards the turnpikes, laws for their exappropriation are quite within the limits of the powers lodged in the States, due compensation being made therefor. Purchased on the basis of their value as sources of revenue to their stock-holders, the actual expense of the transaction would be nothing, for the result would be merely the transfer of values from private to public account. There would be, however, actual profit to the people arising from the saving in the cost of gathering the dues from those who use the roads. It may be noted that the aggregate pay of the toll-gate keepers of this country probably amounts to some million dollars a year.

As fast as a State commission becomes possessed of ways they should be rebuilt in whatever manner the conditions demand, but in all cases the aim should be to secure enduring constructions, so that the expense of maintaining the ways may be reduced to the least possible cost per mile, for the reason that in this country the cost of maintenance is likely to be the main hindrance to the motives that lead to the extension of good roads.

METHODS OF INSPECTION

The inspection of roads has of late become much easier because of the invention of the bicycle. This instrument not only provides swift transportation, but by its motion it indicates to the rider in a very effective manner the condition of the way it traverses. By its use the surveyors of highways may be able to do their work more swiftly than of old, keeping charge of larger numbers of men and more miles of way. With this or other means of travel the rule should be that the whole of each road should be traversed on the average once each month from the time it is opened to the first reconstruction, and when it begins to show signs of wearing as often as every ten days. Experience with the Massachusetts State roads indicates that for the first year or two a well-built road, subjected only to the stress of agricultural traffic, is not likely to require any repairs; yet even on these well-built ways roadmen could advantageously be kept employed, to the number of one man to each five to ten miles of way, in seeding or sodding the slopes, or in caring for the trees. When the time comes for partial repairs it is likely that one man will be required on the average for each four miles of way.

An alternative to the steadfast employment of section men is an arrangement with the town or city authorities whereby they shall, on notice from the surveyor of the board, proceed to make, at the cost of the State, such repairs

as may be required. This arrangement would have the
advantage that it would serve to train the local superintendents of streets in the task of keeping ways in high
condition, with the probable effect that their work on the
local roads under their charge would be much improved.
In other words, the same educational intent which has led
to the plan of making contracts for the construction of
the Massachusetts State roads with the authorities of the
towns in which they lie would be extended to the task of
maintenance. This method is well worth trial, even if the
cost be somewhat greater than that which would be incurred by having the whole matter directly in the hands
of the State agents, for the reason that the most important
problem now before the friends of good roads is as to the
means by which a knowledge of the methods of building
roads can be disseminated among our people.

So far in this country, even where good roads have been
constructed, there has been, save in rare and local instances,
no system in the method of keeping them in good condition. They have been allowed to fall out of repair and to
attain a state of unserviceableness before they have received
any attention. Where the way has come to be nearly
useless it has been reconstructed. This method of action
is as reasonable in the case of a road as it would be in the
case of a house. We all know what to think of a man who
will allow his house to rot down because of the lack of a
water-tight roof, and who prefers to repair the edifice with
a new building; but we have not learned to see that like
condemnation fits in the case of a highway. To clear the
minds of men concerning this matter is worth much in
the way of experiments such as are here recommended
in the project for the repairing of State roads.

As soon as the stage of local repairs is attained the

question arises as to the source of supply of broken stone to be used on the roads. Where there is a crusher in the immediate vicinity the materials can easily be had, but where, as in many cases, the road has been built by a crusher which was temporarily established it may not be easy to obtain them. In this case resort may frequently be had to hand-breaking. Care should in these cases be taken that the stone used in repairs is of the same quality as that of the original construction. Any considerable difference in the nature of the materials, as regards resistance to wear, will be sure to bring trouble to the roadmaster.

REMOVAL OF SNOW

In caring for roads in the northern part of the United States much of the expense goes to clearing away snow. This matter is of such importance that in many districts it needs to be considered in the original project of the road, so that cuts in positions where the snow is likely to drift into them may be avoided. So difficult has it seemed to the Massachusetts Highway Commission adequately to care for the road in this respect by State employees that the board has asked the legislature to require the cities and towns in whose limits the obstructed sections may be to do this part of the service of their own instance and at their own cost. In the northern part of this country, where the people are accustomed to use sleighs or sleds, it is not demanded that the snow be removed from the way, as it is the custom to do in France and other regions where snow is exceptional, but only to "break out" the drifts. This is usually done by the use of wooden shovels or by snow-plows.

NEED OF ELABORATING METHODS

In closing these notes on the methods of caring for roads, it may be said that it will not do to trust much to foreign experience for guidance. We must work out our own methods—perhaps several, fitted to local conditions—of doing the work in a manner to meet the needs. In this task we must, however, bear in mind the fact that the means employed must include skilful supervision of the ways, experience of the same order as guided in their construction, and an immediate remedying of every defect as soon as it is discovered. Each method proposed may fairly be criticised as regards the extent to which it provides for these needs. It may furthermore be noted that in this as in other innovative work it is desirable to do as little violence to existing interests and prejudices as is possible. Therefore, so far as the supervision by the State board can be made to control the work of existing local authorities, that course will be better than the alternative method of having an administration of these roads entirely separated from that which cares for the local ways. There are evident difficulties in this combined action of State and town authorities which may in the end make it impossible to continue the relation, but the advantages are so great that the experiment should be essayed.

On endeavoring to combine local and State action in the management of country roads, care needs to be taken to adapt the method to the needs of the commonwealth in which it is to be applied. Thus, in Massachusetts, where the area is small, the conditions for administration favorable, and the people accustomed to the interference of State boards in their local affairs, it is easy for a commission to deal with the well-organized town authorities.   In the case of States of large area and relatively weak local governments, the task is one of much greater difficulty ; in such conditions it may be best not to undertake the management of the whole field altogether by one board, but rather to divide it into two or more districts.   Where, as is likely to be the case, it seems best to have all the State highway work controlled by a single commission, the roads of the several districts may well be managed by separate engineers, the work being supervised by one of its members.

Breaking stone for a German country road.

CHAPTER X

MACHINES USED IN ROAD-MAKING

Road machines. Plows and wheeled scrapers. Use of explosives.
Stone-breakers. Permanent and movable plants. Road-rollers.
Weight of power-rollers. Watering-carts. Municipal owner-
ship of road-building machinery

ALTHOUGH the variety of modern inventions for use
in the construction and repair of highways is not
great, at least as compared with the number relating to
other tasks of similar economic value, the importance of
these machines is such as to call for some account of
them in this work. First of all, we may note the group
of somewhat complicated engines which are designed to
save labor in repairing ordinary "dirt roads," i.e., those
which have no kind of pavement whatever, but which trust
to compacted soil or to the hard-pan which underlies it for
the support of the wheels and hoofs.

ROAD MACHINES

The road machine is, with the exception of the crusher,
the only contribution of importance which American inventors have made to the apparatus used in highway work.

There is a considerable variety in the form of this machine and in the details of its structure; but in general it consists of a set of scrapers, in part designed to level off the irregularities of the ruts, but in larger measure to scrape out the contents of the side ditches and to cast this matter into the middle of the road. Although there is a superficial look of repair given to a highway by the action of this apparatus which to the untrained eye is very attractive, it is rare indeed that anything more valuable than mere show is obtained from it. The obliteration of the ruts on a dirt road is a temporary convenience. The depressions, being filled with soft materials, are likely to be remade by the next passing of laden vehicles during wet weather, so that the wheels have once more to cut and clear their way down to the hard-pan level which they found or formed, which task they will in all cases speedily do.

The plan of sweeping the contents of the ditches into the middle of the road, so essential in these engines, is utterly irrational and mischievous. All the matter which normally finds its way from the road to the gutters is a good riddance. It usually consists of clay mixed with horse droppings, and is to a great extent mingled with decayed leaves. When returned to the road and wetted it becomes mere mud, which fortunately may work back to the gutters in the next heavy rains. The only thing that can be said in favor of the ordinary road machine is that it provides cheap though temporary ditches which serve for a short time to drain the water from beneath the road-bed. The relief, it is true, is of brief duration. These scrapings of the ditches also serve to remove the growth of vegetation which may have obstructed the flow of the water toward the natural watercourses. These

12

results should be accomplished, as they may be at no great
expense, by the use of ordinary tools, the materials from
the ditches being cast away from the road and not upon it.

Coming now to the really serviceable engines which are
employed in road-making, we may note very briefly the
combined use of the plow and wheeled scraper, which in
their modern methods of employment have done so much
to cheapen the excavation work in this and other constructions. Plows for this service need generally to be made
of peculiar strength, arranged to break up the firmest
earth which can be moved at all by such instruments. It
is likely that in time steam-power will be used for this
service, being applied either by means of cables or by
traction engines. An engine of this description might
easily be devised which would serve, when laden with water
or stone, as a road-roller.

No considerable task of grading can advantageously be
undertaken without the use of the wheeled scraper, a contrivance by which earth lifted into a scoop by the use of
the ordinary sliding scraper can be elevated from the
ground, so that it is no longer dragged over the surface,
but is borne on wheels. With earth broken up by the plow
or by dynamite, excavation can be done by means of this
instrument at a small portion of the cost incurred in the
slower method of pick and shovel.

USE OF EXPLOSIVES

When the cuts are to be made in what is ordinarily
termed hard-pan, i.e., very firm clay or other material of

AXEN ROAD, LAKE OF THE FOUR CANTONS, SWITZERLAND.

the same consistency, too compact to admit of being broken up by the plow, the use of small charges of dynamite is exceedingly helpful in cheapening the expense of the work. Usually this explosive should be applied in half-pound cartridges in drilled holes considerably longer than the tubes. If the conditions admit, these charges should be placed a few feet back from a working face of the cut, at intervals, and with a depth below the surface which experience with the particular ground may indicate as most suitable; the firing being done by a small electrical battery such as is obtained at a small cost from merchants who furnish contractors' supplies. This method of exploding the dynamite insures the simultaneous action of all the charges; moreover, it diminishes the danger arising from the use of such powder. It is hardly necessary to say that dynamite should be applied by those persons only who have had a training in its use. With fit precautions this agent can be made to serve the needs with very little danger to the operators. It should, however, be altogether in the charge of an experienced person, who should see to the storage of the material, place and give the charges, and make sure that all the cartridges have been exploded at each time of firing. One of the most serious kinds of accident arising from this dangerous substance comes about from overlooking an unexploded charge, which may be fired by some chance blow of a pick in the hands of a workman.

The modern system of carrying materials by means of wire ropes on which buckets are hung is occasionally applicable in removing excavated materials to points where they are to be used in filling or are to be cast aside. The conditions of such application are, however, so rare in highway work that this process demands no more than a mention.

STONE-BREAKERS

The most important piece of road machinery now in use is the stone-breaker or crusher. Its application to the service of road-building has not only greatly cheapened the expense of breaking stone, but has insured that the work may be done under conditions which permit the bits to be assorted according to size, a feature which, as before noted, is of very great advantage. It has enabled the road-master to build ways at much less expense in terms of labor than he could with hand-broken stone, and these ways are free from the evils which are due to the ill assortment of sizes in the layers of the road.

As regards the method of their action, there are two principal types of the crushers now in use. In the older of these, which follows on the plan laid down in the original invention of Blake, the work of breaking is done by means of very stout jaw-plates which open and close after the manner of nut-crackers, so that they may take in large masses of stone, which, by successive fracturings as they descend into the jaws, are crushed to the size which for its maximum is determined by the distance apart of the plates at the lower margins of the jaws. Another method of effecting the crushing is by means of an arrangement somewhat like a large steel mortar, in which a massive pestle swings around so as to bring a pinching pressure to bear in a circular movement against the rim. There is probably not much difference in the efficiency of these two types of machines. Though varying in appearance, they involve the same principles of action. As yet the comparison between them has not been sufficiently complete to determine their relative economic value. Into this determination there enter many considerations besides that of daily product with a given expenditure of power. Among these may be noted the cost of repairs and the expense of moving the crusher plant from point to point and installing it in its successive positions.

More important than the particular method by which the power is applied to the stone-breaking is the arrangement of the apparatus so that it may do its work cheaply and effectively. The plant should be furnished with rotary screens so placed as to bring about a proper sizing of the product. This product, according to sizes, should go to bins or pockets so arranged as to discharge directly into the carts or cars which are to convey it to its destination. These bins should be of such dimensions that they will hold all the product of the crusher during the interval when the removal cannot be effected. There should also be carriers for the bits which pass through the crushing jaws and are yet so large that they are rejected by the sizing-screens, so arranged that these bits may be returned to the breaking apparatus. Where the stone yields an excess of "fines," as is often the case, it is also desirable to have some automatic system of conveying this material to a waste heap. By these self-acting contrivances the cost of moving the broken stone by hand is avoided. The importance of this is seen when we remember that the expense of moving such material by hand is likely to be from eight to twelve cents per ton, so that one such carriage may amount to about one tenth of the cost of the product of the crusher.

It is also very important to have the stone-breaking plant so arranged that the wagons which bring the stone from the quarry or from the fields should have ready access to the dumping-platform beside the crusher, and

12*

that the conveyances which take it away should have a like access to the mouths of the bins. All these labor-saving arrangements require a very careful adjustment of the plant with reference to the station which it occupies. It is, moreover, necessary that there shall be good wheel-ways maintained about the machine, which is often not an easy matter to effect. In fact, when a crusher is moved from place to place along a road, each "set-up" being made, say, at intervals of two miles, it is practically impossible in most instances to obtain even a tolerable adjustment of the needs of water-supply, cheap stone for crushing, and ready access to the plant.

PERMANENT AND MOVABLE PLANTS

The foregoing considerations bring us to the point where we must briefly discuss the relative advantages of permanently established crushing plants so placed that they may be immediately adjacent to quarries of well-selected stone, as compared with plants which are intended to be moved about in the manner above indicated. In favor of the fixed plant at the source of stone-supply it may be said that it has the advantage of yielding a uniform high-grade product from an establishment which can be elaborately organized for the work. The quarries may be operated by means of steam-cranes and drills, and the stone conveyed to the breaker by rail carriages. The breaker, not being limited in weight, as is the case with movable plants, may be of larger size of jaws and more massive structure, thus avoiding, on the one hand, the cost of sledging the stone so as to make it fit the machine, and, on the other, the excessive expense of repairs. Where the stone is delivered to railway cars or to boats there is no question as to the state of the ways it has to traverse, as is the case with the movable apparatus at each of the roadside stations. Against these advantages is to be set the cost of carriage of the material from the permanent establishment to the point where it is to be used. This is in all cases expensive, as it involves railway or boat transportation and usually at least one unloading from the conveyances by hand.

Nevertheless in a region which is much intersected with rail or water ways there are likely to be many roads which can be more conveniently supplied with broken stone from fixed crushers than from those which are moved from station to station. In Massachusetts the average distance of railway stations from existing permanent stone-breaking establishments is not far from forty miles. The actual cost of railway carriage per ton over this distance is probably not more than half a cent per mile, but in no case is it to be reckoned at more than two and a half times that sum. It is, however, to be noted that, with one exception, none of the railways in Massachusetts will move stone for any distance, however short, for less than fifty cents a ton per mile. If the railway authorities could be brought to see that every wagon road over which freight is brought to them is in effect an extension of their own lines and therefore deserving of their help, we might hope that they might adopt the plan of carrying road materials at cost. If this were done the advantage of permanent crushers in the New England States would be so great that it is doubtful if the movable forms of such apparatus would be used, except where the wagon distance from a railway station to the point where the broken stone had to be used exceeded three miles.

Much attention has been directed to making crushers more movable than they are at present, so that the cost, on the average under existing conditions of from one to two hundred dollars, incurred in moving the machinery over intervals of, say, two miles, may be lessened. There are, however, serious hindrances in the way of success in this endeavor, for if the machinery is made lighter it is necessarily frailer, the costs of repairs are greater, and the labor-saving devices prove less effective. There would probably be no serious mechanical difficulty encountered in placing the crusher itself on wheels, so that its weight could be moved from point to point as easily as any other wagon with a load of, say, five tons upon it. But this apparatus alone would produce unsized stone, most objectionable material to use on a road, and it would need to have the rough stone and water drawn to it at a cost which would be likely quite to offset the apparent advantage of the arrangement. It therefore seems, in the present state of the art, much the best to adhere to the existing plan of setting up a substantial crusher upon its subsidiary apparatus at convenient intervals, usually about two miles apart where field or quarry stone may be obtained as well as water, and hauling the product to the point where the material is to be placed on the road. No account need be taken of the wearing of the road which will be brought about by the wagons conveying the stone which pass over it. Experience shows that this action at the time when the road is building not only does no harm, but it actually benefits the work, provided the teams be made to use all parts of the hardened way.

In selecting a crusher attention should be paid to the size of the stone which can be fed into its jaws. The smaller this size the more the work with the sledge in the preliminary breaking of the stone will be required. If the opening be of the smaller gages the cost of this sledge-work may rise to twenty cents or more per ton. A difference of an inch in the diameter of the stone which can be taken into the jaws makes a notable reduction in this element of cost. To determine the needed size of a crusher requires a careful study of the average diameter of the field stone, or of that of the masses of rock which are broken from a quarry under the action of the blast.

It is a matter of much importance to have the available product of the crusher amount on the average to about one hundred tons per day, this being the quantity which can be conveniently treated by an ordinary roller. Moreover, the operating expenses of the machinery are but little greater with this amount of production than they would be with one half the above-indicated product. It is well to have a distinct contract with those who furnish the machine that it shall be able to maintain this rate of breaking. To maintain a uniform rate of production spare parts of those elements of the mechanism which are likely to become broken should be kept in stock. These parts are likely often to give way in the exceeding strains to which they are subjected, so that in the end there is but little expense incurred by making this provision. Speed in repairing saves much, for when the crusher is broken the roller is necessarily idle and the teams without work. It is indeed true economy to keep duplicates of every element in the machine which cannot be readily repaired by an ordinary mechanic. It would in any case be better to have the whole apparatus in duplicate and ready for action rather than encounter the delays which happen when, in case of breakage, resort must be had to machine-shops, it may be, hundreds of miles away.

ROAD-ROLLERS

Even more important than the crusher with its sizing-drums, which serve to divide the stone into the grades before mentioned, is the roller which has of late years come into use in the construction of high-grade Macadam roads. In the early stages of the art of building roads of broken stone it was necessarily the custom to trust the compacting of the mass to the treading of wheels and hoofs. In the course of time these instruments will well drive the stone together, forming enough dust in the process to bind the mass. The process is slow and in two important respects costly. Before the stone is compacted the tax on the draught animals is so great that the road is likely to be less useful than an ordinary earth way. This impeded state of the way is apt to last, when the traffic is not great, for some months. During the period when the road is "coming down" the stones are driven about by the wheels and hoofs in such a manner that they often become very much worn. They in all cases lose the sharp angles which are so useful in holding them together when they are promptly settled into their destined bed places as by the use of the roller. The loss of bulk which takes place in this process has not been determined, but it probably amounts in some cases, especially where limestone is used, to as much as one fifth of the mass of broken stone which is placed on the road. The above-noted primitive method of management may still be observed on the roads built of broken limestone in Kentucky, and in other parts of this country where the ways are in the hands of rural corporations who have their profits from the toll-gates, and whose administration is as medieval as the spirit of their charters.

Observing that the wheels served to " bring down " the road, it was natural to seek the same result by the use of iron rollers.   These were at first made on the pattern of the ancient wooden tool used in breaking the clods of the fields.   It was soon found desirable to construct the roller of two or more independently moving sections ; this change finally led to the invention of the disk-roller, made of many separate, closely adjacent wheels, each with a certain free play on the axis upon which it revolved.   It was also found advantageous to provide means by which the weight of the machine could be increased when the way became partly smoothed, so that it could be more easily drawn in the first passages over the rough surface.   This additional weight was obtained by filling boxes in the frame with stone.   The horse-roller was exceedingly useful.   It changed the practice in construction in an important way.   Where the treading action was left to the vehicles it was the custom, the best which under the conditions could have been adopted, to allow the broken stone to work down the lower layers of the road, forming in a costly way a rude foundation for the superstructure of the way.   Into the depressions of this bottom layer there was then filled stone in the manner followed in the process of repairs.   The iron roller drawn by horses enabled the road-master to fit the structure for final use at once.   The difficulty with these machines was twofold.   They were not heavy enough to compress the stone to the needed extent, so that the bed lacked the solidity we now can give it.   The feet of the numerous horses required to draw the machine pulled out the stones to such an extent, especially in ascending slopes, that the work was slowly accomplished and was attended by much rounding of the broken stone.

In all modern work which is done in the best manner of

the highwayman's (we use this excellent word in a bettered modern sense) art, the steam-power roller has taken the place of that drawn by horses. The advantage of the modern instrument in all except the purchase price is very great, and the cost of compressing a given amount of stone by it is so much less than that incurred by the use of the ancient instrument that the price should in most cases not be reckoned. There are no sufficient data for estimating with accuracy the relative costs of building Macadam roads with or without power-rollers, but they are probably at least two to one in favor of the steam-machine. In this regard the firmer road-bed given by the modern engine would alone warrant its use.

In its existing form the steam-roller appears to have attained very nearly to its possible development. Its main defect in the usual pattern consists in the small size of the drum-like wheels which are placed in front of the middle of the frame. Because of their limited diameter these wheels, after the manner of the small fore wheels of many wagons, tend to push the stones in front of them rather than to tread them down. It would be desirable to contrive a model for this engine which would permit of an equal amount of weight on four wheels, each of like and large diameter, the larger the better, so that the effect would be the same in all parts of the path of the machine.

It would also be a decided advantage if the steam-engines of the roller could be replaced by some form of hot-air motor. At present the inconvenience of furnishing these machines with water is considerable, and, as the weight of the apparatus is not objectionable, there is no evident obstacle to the use of this means of applying the energy obtained from the fuel.

WEIGHT OF POWER-ROLLERS

There has been much discussion as to the most suitable weight to be given to rollers. The first experience seemed to indicate that this should be not less than about twenty tons. It is still open to debate whether for the hardest kind of stone, such as the firmest trap, it is not the better for treatment with a machine of this order; but for materials of average resistance, and particularly for the softer kinds of stone, the pressure which this size of roller applies is clearly excessive. The broken stone is evidently subjected to a further breaking, a process which is very undesirable. Moreover, where the material is suddenly jammed down without much movement, the junction of dust between the adjacent faces of the bits does not take place in sufficient amount to effect the needed cementation. A compact mass is formed, but the material to bind the bits together is lacking. Furthermore, a roller of this weight tries the foundations over which it moves. It is very difficult to find bridges strong enough to sustain with safety the passage of the heavier rollers, and thus their movements from one piece of work to another are restricted.

There is at present a tendency to diminish the weight of rollers. Few are made of greater weight than fifteen tons. Many now in service are but twelve tons. Those of ten tons' weight are likely to come into use. As the weight of the roller is diminished the time required will have to be the longer because of the increased number of passages which the machine has to make in order to do a given amount of work. Therefore there is a point beyond which the reduction of weight will be inadvisable. This minimum limit is probably about fifteen tons for use on the harder traps and twelve tons on the softer or more brittle materials, such as ordinary field stone and limestone, or the cherts. The last-named group of rocks, because of the ease with which the fragments are fractured, are likely to be ground to powder by the heavier forms of rollers. It would be interesting to try the experiment of constructing a roller after the fashion of a traction engine, having a minimum weight of, say, seven tons, with arrangements so that it could be loaded with metal bars or stone up to a maximum of twelve tons. In its lighter form this machine might be used to draw the heavy plows used in grading the road, and in some instances, perhaps, to draw the wagons containing the broken stone from the crusher to the point where the construction is going on. In its unloaded state it would also serve to roll the stone used in repairing the depressions in the way, in which task it is undesirable to have the stone compacted by the heavier weight applied to the material. A roller of this light weight would be able to traverse ordinary roads and bridges, in going from one piece of work to another, more speedily and with much less risk than one of double the weight. If it were found best to give the increased weight by means of iron or lead, a wagon drawn by the traction engine would serve for the conveyance of the material from place to place.

Although the power-roller greatly facilitates the process of building roads where the pavement is formed of broken stone, it must not be supposed that no disadvantages are to be found in its use. Those who have carefully watched the effect of this machine may have noticed that, however perfectly the stone may have been pressed down so as to form a solid-looking bed, the bits are likely to "pick up"

ROADS AT LUNGERNTHAL, SWITZERLAND.

under the action of the wheels and hoofs, so that the road, especially if dry weather follows its construction, appears to be going to pieces. After a time, however, the wheels serve to draw the fragments together so that the union becomes firmer. It is indeed a mistake to suppose that the process of adjusting the broken stones to one another is complete when the work of the roller is done. At that stage of the construction the pieces have been forced into a preliminary and partial adjustment; the action is completed when the pressure from the wheels of the carriages and the blows from the horses' feet have pushed the stones this way and that until they are well fitted to one another.

It seems a worthy task for our inventors to contrive some form of roller which will combine the immediate smoothing effect which is now attained with the result which, as above noted, is brought about by the action of the ordinary use of the way. It seems likely that the end might be attained by corrugating the surface of the drums or broad wheels of the machine, so that they would combine a shearing movement with the simple downward pressure which they now apply.

It has been suggested that some combination of roller and crusher might be devised which should be so constructed that one engine and boiler might serve as a source of power for both instruments. The criticism on this suggestion made by my colleague, Mr. W. E. McClintock, engineer member of the Massachusetts Highway Commission, shows clearly that the project offers no advantages. This criticism is in effect that the plan proposes to use the high-grade, costly engine of the locomotive type, which drives the roller, to do work which can be better accomplished by a much cheaper engine at the crusher. Moreover, it may be said that any system of working which involves the crushing at one time and rolling at another time would necessarily lead to two or more handlings of the broken stone by means of shovels, a process which would cause an expense of twenty cents or more on the product of the crusher.

WATERING-CARTS

There remains to be noted the apparently simple device of the watering-cart which is used in watering the broken stone at the stage of the construction where the "fines" are placed on the surface to be rolled into the bed. It is advisable that this machine should be something better than the rude contrivance which is often made to serve the purpose. In the first place, the tank should be as large as two horses can easily draw over the grades which have to be passed. It should be mounted on a carriage with broad treadwheels of considerable diameter. The tires should not be less than five inches in width; they should be rounded at the edges. This arrangement is the more necessary for the reason that the load has often to pass over soft ground as well as over pavement which is imperfectly consolidated. The weight of the vehicle often exceeds four tons, which pressure on the lower layer of macadam may serve to make distinct ruts. The discharge of the water should be controllable so that it may be sent forth from either half of the spray-pipes; the exit should be as a spray, and not as a broad, splashing sheet, which is likely to pour forth so much that the dust may be washed out. It is essential to water the road for rolling in a much more complete manner as a preliminary to rolling than where, as in the ordinary use of the watering-cart, the aim is to dampen the dust. Enough water needs to be applied to afford a surplus beyond what is required to convert the dust into mud.   It is necessary to wash the dust down into the crevices.   These objects will best be attained, not by one profuse discharge of water, but by two or three movements of the cart over the same area.

Where the care of the road is of a high order, the "fines" worn from the road should from time to time be removed, no more being allowed to rest upon the surface than is needed to furnish the cementing material which should be continually allowed to work down between the crevices of the broken stone.   In practice the choice is to remove this surplus of ground rock in the state of dust or in that of mud, i.e., by sweeping or by scraping.   On the whole, the preference may be given to sweeping, for the reason that the weight of the material to be removed is least in the dry state, and the wearing action of brushes of any form is rather less than that of scraping tools.   The European method, seldom employed in this country, is to sweep with brush-brooms; here the rotary horse-power sweeper is generally employed.   There is considerable diversity in the pattern of these machines.   All those employed seem to be moderately effective.   Hand-scrapers of the best form have the edge made of india-rubber, which is less apt to pull out stones than wood or metal edges.   The horse-power scrapers are best arranged so that a number of distinct blades operate separately, though in the same time, so that they adjust themselves to the irregularities of the road.

Although the variety of machines which have been contrived to cheapen or better the processes of making or caring for roads is not great, they meet the needs of the art in a tolerably effective way.   The difficulty is that they are, at least as regards the most important, so costly, as

13

regards both the purchase price and the expense of operation, that they are often beyond the means of small municipalities. Stone-breaking plants properly arranged with screens and bins may be reckoned as costing from $1600 to $2700; rollers from $2200 to $3500; serviceable sprinkling-carts from $325 to $400; so that an outfit for road-building, including only the indispensable apparatus, is likely to cost from $4125 to $6600. It is necessary also to reckon that a competent engineer is needed for the crushing machinery and another for the roller. Such men, if well trained, are dear; they are dearer yet if untrained; their labor should be worth not less than three dollars per day. The cost for repairs and depreciation of machines of this description used in the best modern practice is, in proportion to the number of days of service each year, quite large.

MUNICIPAL OWNERSHIP OF ROAD-BUILDING MACHINERY

The question as to the most advantageous method of obtaining possession of road-building machinery, whether by hiring it for each particular service or by purchasing it outright, is often asked by municipal authorities. This is not an easy inquiry to answer. In general it may be said that where the assessed value of a town on the usual basis of estimation amounts to as much as one million dollars, and where the roads which should be built of broken stone exceed twenty-five miles in length, it is most advantageous for the community to own a complete outfit for road-making. Where, however, the valuation is less than that designated, it is, perhaps, in most cases better to have the work done by contractors who own the machines. Such men are becoming plenty in the richer parts of this country; they may be expected, because of the greater cheapness of contract work, to accomplish constructions at but little, if anything, more than they will cost if done by day's labor employed by town officers. The distinct advantage arising from the possession of road-building apparatus is found in the inducement which it gives to do work each year, so that the community is likely to go much further and faster in road-building than where it acts through occasional contracts.

It should be noted that the apparatus for road-making is apt to suffer much depreciation due to careless exposure to the action of the weather, especially during the long winter season when it is not in use. This use should always be discontinued during the time when the ground is likely to be frozen. In the Northern States of this country the fit time for road-building is from May 1 to November 1, or but half the year. In the winter season the machines should be carefully housed; even the crusher, or at least all the working parts of it, should be protected from the weather. Such care will much prolong the time of service of these instruments.

Where individual communities are not rich enough to own a set of road-building machines, several lying adjacent to one another may well coöperate in the undertaking. It may be estimated that with due preparation a mile of road, after grading, can be built in six weeks, making due allowance for the time required for the necessary moving of the plant. Four coöperating towns owning the machinery in common could each build one mile of road every year. In this manner small places could have the advantage of using the road-building plant without the undue expense involved in its purchase and maintenance.

CHAPTER XI

THE COST OF ROAD-BUILDING—SIDEWALKS—PARAPETS —CITY STREETS

Cost of Massachusetts State roads. Cost of gravel roads. Sidewalks. Guard-rails and parapets. The pavements of city streets. Principal varieties of pavement. Difficulties arising from tramways. Classification of streets as regards traffic

THE costs of road-making and repairing have, unfortunately, been very inadequately determined. These data have not been gathered for this country. The results of European studies are not applicable to this part of the world, for the reason that the rates of wages and the efficiency of labor are not the same in diverse parts of the world. Moreover, the greater part of the reckonings were made before the power stone-breakers and other labor-saving devices came into use. On this account the following statements concerning the cost of making gravel and macadam ways should be taken as rough estimates which will serve no more than to give an approximate idea of the expense of such undertakings.

In most cases the expenses of building a road may be divided into two principal groups: those chargeable to the preparation of the road-bed, such as grading, drains, culverts, and bridges, the cost of which may vary indefinitely; and those to be reckoned as pertaining to the hardened portion of the way. This latter group of costs can usually be clearly reckoned. In an appendix will be found a table extracted from the annual report of the Massachusetts Highway Commission for 1895, which indicates the cost of various items in road-building. The original publication gives the statement for expenses for a much greater number of items than are here included. A selection has been made of those elements of expense which are most indicative as to the matter of cost.

As before remarked, the Massachusetts commission has followed the plan of taking the worst portions of those roads which were the most important and the most ill conditioned of the ways in each district. This policy has naturally resulted in making the average expense per mile of construction considerably greater, perhaps one third to one half more than would have been incurred if the selection had represented the average condition of the ways in the commonwealth which are naturally to be classed as State roads. Yet, as a similar policy should be followed in all such work, it is not unreasonable to suppose that the average cost of the work done by the commission is about that which will, under like conditions, be at the outset incurred in building first-class macadamized ways in any part of this country.

COST OF MASSACHUSETTS STATE ROADS

The average cost of the ways already built by the Massachusetts Highway Commission, including various incidental expenses connected with the administration of the work and also certain charges for seeding down slopes, tree-planting, watering-troughs, etc., is about nine thousand dollars per mile. The range in cost is from three

13*

thousand in the case of a very well-built gravel road to about twenty thousand dollars in the case of each of two macadamized ways where the grades or foundations presented abnormal difficulties which led to great expense.

The cost of the hardened pavement has varied greatly, but at nothing like the rate of the foundation work. The variety in expense, as will be noted in the tables given in an appendix, has been due principally to the range in the cost of the broken stone itself. Where the material was readily accessible field stone has been used. The cost of this work has, in cases, been less than one half that incurred by the use of imported material. The average cost of this broken stone in place in the pavement has been not far from a dollar and a half per ton where the material was obtained from the fields, and two dollars where it was taken from distant ledges. The greater part of this imported rock was trap. All of it was brought from permanently established crushers. In practically all cases it was subjected to the costs of rail or water transportation.

The variation in width, and other conditions of the hardened way which exist on these roads, make it impossible to give any accurate statement as to the average cost of the macadamized portion. An inspection of the results shows, however, that this cost is not far from five thousand dollars per mile, including such Telford foundations as are likely to be required on a total length of, say, fifty miles of way. With a thorough organization of the business, such as is always possible in work which is continued for a number of years, it may be possible to cheapen the cost of this portion of the work to the amount of about one fifth of the above-named sum, but at the present price of labor and materials it will probably require very good management to attain this end. In the conditions which

A STONY ROAD IN BAD REPAIR.

Codman street, Dorchester, Massachusetts, near stable of the Municipal Paving Department.

prevail in New England it must be reckoned that the main rural ways of communication will require an average expenditure of from seven to nine thousand dollars per mile to bring them into thoroughly good and serviceable condition. This sum will not include the cost of the greater bridges, say those with more than twenty feet of waterway, nor the expense of tree-planting, watering-troughs, or other accessories of the construction.

It will be noted that these estimates of cost are much above those which are stated to have been incurred in other parts of the country in building macadamized roads. Reckonings have been given, which are claimed to be based on practical experience, which elsewhere show these costs to be as low as twenty-five hundred dollars per mile, some, indeed, at the preposterous price of nine hundred dollars per mile. The answer to this criticism as to the costs of Massachusetts roads, which are about the same whether built by the State commission or by municipalities on their own account, is that the cheaper roads are so poorly built that they are in the end the more expensive. It is quite possible, by spreading rubble-stone upon a road with no proper care to the foundations or to drainage, bringing these stones to something like a level by means of hammers, and then placing a layer of two or three inches of unsized stone from a crusher upon the surface, to make a way which has a width of, say, twelve feet which will not cost, under favorable conditions, more than two thousand dollars per mile; but in a short time such a road will wear out. Its life is not likely to be for more than three or four years, and the first mending will require " general repairs " at which no part of the material used in the way may be found serviceable. On the other hand, the properly built way will be likely, with the same amount of use, to wear, with local and little costly mending, for fifteen years. It may, indeed, require nothing more than trifling care for the first five years of its service. When the original thickness has worn down to a coating of, say, four inches in depth, this can be picked up and reincorporated in the new pavement. Under these conditions the dearer construction may in the end prove to be so much more economical than the cheaper that it would be folly not to make use of it. Taking account of the difference in the quality of the two kinds of roads, of the expense of repairs and reconstructions, and the embarrassment of traffic arising from the rebuildings, computation shows that a way costing in the first instance two thousand dollars per mile may prove, principal and interest reckoned, far more expensive than if it had been built in an enduring manner at the outset at a cost of, say, nine thousand dollars per mile. A private person may, under certain conditions, find a profit in building such temporary ways, but the essential difference between an individual life and that of a community is that the latter must reckon on an indefinite term of action.

COST OF GRAVEL ROADS

Where gravel of good quality for road-building can be found near the road, the first cost of a hardened way, so far as the pavement is concerned, need not exceed one half, and may not amount to one fourth, that required for a road made of broken stone. This lessened cost is due to the fact that the expense of breaking, rolling, and watering is spared. Moreover, the distance to which the material has to be hauled is often less. It is, however, to be noted that graveled ways are certain to prove somewhat more troublesome in regard to repairs, and are likely in a term of years to be more costly to keep in order than those which are macadamized. Nevertheless, as before remarked, in determining as to the manner in which a country road is to be improved, care should be taken to ascertain whether the conditions do not admit of its being covered with this gravel. If the need of the traffic can be met by such a road and the material for use is at hand, the capitalized cost, including the expenses of repairs made with all desirable frequency and care, probably need not be greater than one half that of the dearer form of construction with broken stone.

The peculiar advantage of gravel roads, as regards their cost, is that they require no investment of money in rollers, crushers, or other costly machinery. There is, it is true, a certain advantage in rolling the surface of the bed before the gravel is applied in order to bring it into shape for use, but this is by no means necessary. All the needed compacting can be left to the carriage-wheels, or may be in part effected by horse-rollers. No use of the steam-roller within the practicable limits of its continuance will cause gravel to "come down" in the manner of ordinary broken stone. This pressure, in the present state of the art, had best be left to the action of the vehicles. Therefore once again it is urged that, while the pavement of broken stone is in most instances the best and in many cases is indispensable, the fitness of any road for the service of a community should be made the subject of careful inquiry before adopting the costlier method of construction.

SIDEWALKS

Too little attention is given in this country to the footways alongside of the carriage roads. In the more cultivated portions of Europe provision for them is made; in many districts such ways, miles in length, are well cared for. In many districts, such as parts of the Alps, where it would not be economical to construct roads for wheels, foot-paths are carefully maintained. Not infrequently in the Old World the ancient tracks once traversed by pack-trains remain as paths for men. Here and there the traveler may find dwellings, and even small hamlets, which have never been approached by a wheeled vehicle. Similar conditions exist in many of the remote valleys of the southern Appalachians, where a bridle-path affords the only means of approach to considerable settlements.

As the blow of the human foot is light, sidewalks, except in densely settled places, are easily kept in repair. They need be of no great width. The main object should be to guard against mud. This can be most effectively accomplished by a construction made in the following described manner. A ditch about one foot in depth should be excavated for the width of the proposed path. This work can generally be done by the plow and ordinary scraper. This should be filled with pebbles, gravel, or broken stone, the coarser material at the bottom. On this foundation there should be a layer of two or three inches of gravel of a quality that will cohere; the dust from the stone-crusher, of which there is generally a surplus, is well fitted for this layer. This surface should, if convenient, be rolled with a horse-roller after watering, but it will serve pretty well without this additional treatment.

When the earth on which the sidewalk is built is not readily permeable by water, drains filled with pebbles or, better, of small clay pipe should be carried to the gutters of the main way at intervals of about one hundred feet. The surface of the walk should rise two or three inches above the level of the ground about it. Thus constructed, the walk may be expected to remain without grassing over and with little effect from the action of frost. It is not likely to need repairing, except, it may be, by the addition of materials to the top coating.

Sidewalks of the type above recommended are suitable for use in communities where, though the houses are much scattered, there is a good deal of passing afoot. Where they can be afforded they should be built along all the main ways leading to school-houses for as far as the children walk. Under favorable conditions such paths can be constructed for four or five hundred dollars per mile. Cheaper, yet serviceable, paths can be made by clearing away the sod or mat of roots, and placing on the surface a layer of gravel or the waste dust from the stone-crusher. This layer should be at least four inches thick.

Where sidewalks are not practicable apart from the main road-bed, as in cuts, it may be desirable to extend the non-metaled shoulder of the way, so that it may be twice as broad as is needed in the interests of the vehicles. This will serve as a foot-path. It is open to the objection that the footman is likely to be in some danger while on the main road.

GUARD-RAILS AND PARAPETS

In proportion as the grades of a road are improved the resulting embankments become sources of danger to vehicles and their occupants. It is therefore necessary to provide barriers sufficiently high and strong to insure protection against the chance of vehicles falling down the slopes. In Europe the usual guard is a stone wall on either side of the embankment down which a vehicle could be precipitated. Such walls, though from their permanence not in the end so costly as they may at first sight seem, are beyond the means of most American communities. Moreover, the depth to which frost penetrates in America makes it more difficult to insure safe foundations for masonry than in the Old World. On this account the usual protection to the sides of the embankments in this country is made by means of guard-rails. As this rail is costly and subjected to decay, it is usually much better, when the height of the embankment is slight, say less than three or four feet, to extend the slope so as to obtain a grade which may be so gentle that serious accidents cannot occur to vehicles which leave the road.

In the Massachusetts State work guard-rails are required to be made as follows : Posts of cedar or other wood which endures well in the soil are set at intervals of ten feet, and one foot in from the edge of the embankment. These posts are planted to the depth of three feet, and project for three feet six inches above the ground. The top of this post is transversely notched, so as to receive one half of a rail four inches square. Half-way down the post it is notched to receive another rail two by six inches in size. These rails, preferably of planed spruce wood, are spiked to the posts. To insure the better preservation of the wood and its visibility in the night-time it is painted with two coats of oil paint of some light color.

It seems likely that for the protection of the slopes from washing or sliding, as well as to guard against the falling of vehicles, it may be well to plant the declivities with some swift-growing species of trees or stout bushes. For this purpose the larch is well adapted because of the speed of its growth and the beauty of its foliage. Under favorable conditions trees two years old, the best size for planting, will in ten years make a dense growth which would afford a better barrier than guard-rails after they had become partly decayed, as they surely would do at the end of that time. By planting the upper row of trees, say, two feet down the slope from the guard-rail and at intervals of ten feet, they would in time serve as posts to which, if desired, rails could be fastened.

THE PAVEMENTS OF CITY STREETS

Although, as before stated, it is not in the purpose of this work to treat of the problems encountered in the construction and management of city streets, it may not be deemed amiss to devote a little space to certain considerations as to this class of ways, especially as to the conditions encountered in constructing and maintaining them in the lesser towns. Where, as in our great cities, the duty of building and caring for the ways is in the hands of able engineers, the presumption is that they are so far competent to meet the difficulties they have to encounter that general advice as to their duties would be misplaced. In the smaller towns, however, the care of the streets is often necessarily confided to men who have had but little experience in highway construction.

PRINCIPAL VARIETIES OF PAVEMENT

At the present time the greater part of the streets of the towns great and small in this country are paved with broken stone. This method of construction is to be commended as on the whole the most satisfactory for the cost incurred, provided the stone which is to be used is of the more enduring kinds, and the travel over it not of a nature to insure its speedy destruction. In most American towns it is the custom to do the conveying of coal and other gross materials in very heavily laden wagons, with the result that on the business streets a covering of any kind of broken stone wears out with great rapidity. This entails ill-conditioned ways or frequently recurring general repairs, with the consequent interruption of travel. In all cases where the traffic is heavy the result, unless the cleaning be frequent and systematic, is very dirty streets. Therefore, except on those ways which have little other traffic than that which is related to dwelling-houses, it is generally impolitic to essay roads paved with broken stone.

Where the amount of wear on a macadamized road exceeds about three quarters of an inch each year, the presumption is that true economy demands a substitution of pavement of some more enduring material. The choice of this harder surface lies between blocks of granite or of trap, brick, or asphalt. Pavements of wooden blocks, once much in vogue in this country and still to a considerable extent used in Europe, appear to be on the whole unfitted to the conditions of American climate. Moreover, experience has shown that pavements of wood often become objectionable from the quantity of foul matters which they absorb, and which make them offensive in the hot American summers.

As between a pavement of stone blocks, of brick, or of asphalt the choice may well be made on the basis of the relative cost of these materials, which varies considerably in different parts of this country. In New England and southward along the Atlantic coast, and in the region adjacent to southwestern Missouri, where good stone for block pavement is obtained, the choice as determined by cost is naturally in favor of stone. In western New York and in the greater part of the Mississippi valley, where good clays for making paving-brick abound, and where fuel for burning is cheap, brick is likely in almost all cases to prove the best resource, except on streets which receive exceedingly hard wear. In the greater part of this country, where asphalt is subject to a large cost for transportation, the expense encountered in using it on roads is likely to make its use wasteful, except under peculiar conditions, which are now to be noted.

In considering the character of a city pavement, attention should be paid to the noise which is caused by the passage of vehicles over it. Where stone blocks are used, owing to the exceeding hardness and ringing nature of the material, and also to the necessary irregularities of the way thus paved, the din produced by the traffic is a serious nuisance. In fact, the infliction arising from this evil is one of the most objectionable features due to this kind of pavement. It appears possible to mitigate the noise from the pavement of stone blocks by having the masses which are used of larger surface and with less intervals at their junctions. There seems no reason, indeed, why it may not be possible, with the modern methods of stone-cutting, to provide a pavement of granitic or gneissic rock which shall be so smooth that it will not resound in any objectionable manner to the blows of the wheels and feet, but as yet such constructions have not been made.

Pavements of brick, when the material is of the best kind and the laying properly done, afford very enduring ways, which can be kept in a more cleanly state than those made of stone blocks, and which are much less noisy. The reason for this is that the depressions, the units of the pavement, are less deep and the materials of less resounding nature.

Asphalt pavements are apparently not well fitted for use

where the traffic is of the heaviest kind, for the reason
that the rate of wearing is, under those circumstances,
great. They are, however, more easily kept clean than
any other pavements. They are much less noisy than the
ways made by either of the methods of construction before
noted, in this regard being only surpassed by wooden
pavements. In regions of excessive heat pavements of this
description soften to such an extent that the wearing is
much accelerated. In dry weather, moreover, unless
carefully watered, they afford an objectionable dust, and
the odor from them, although by no means unwholesome,
is to many persons objectionable.

DIFFICULTIES ARISING FROM TRAMWAYS

The existence of tramways in city streets much limits
the choice of materials to be used in the paving. Where
the traffic is heavy it is very difficult to maintain in good
order the portion of the street lying next the rails. Something of this same difficulty is encountered in the use of
brick, but it may be met in various ways, as by the use of
a narrow strip of block pavement next the railway. In
general, however, where railways traverse narrow streets
the only satisfactory pavement appears to be that made of
stone blocks.

In the immediate neighborhood of great cities there are
likely to be ways which, though they traverse rural districts, are subjected to a traffic so heavy as to demand the
use of block pavements. Where this is the case, as, for
instance, on the main highway between the cities of Boston
and Lynn, Massachusetts, it may be desirable, where possible, to have two diverse kinds of pavement arranged in
strips parallel to each other: that of blocks for the heavily

laden vehicles, and that of macadam for those of a lighter description. The success of such a method, however, would depend upon the efficiency of regulations which would serve to keep the two classes of vehicles apart.

CLASSIFICATION OF STREETS AS REGARDS TRAFFIC

It seems desirable in our cities to follow the plan, so generally adopted in the greater towns of Europe, of separating the freight and pleasure traffic, so that the use of certain roads which may be paved with macadam may be reserved for the lighter class of vehicles, while certain other ways, paved in a more enduring fashion, but in a manner objectionable to those who ride for pleasure, may be kept for freighting purposes. By the institution of such a division the noise arising from the passage of heavily laden vehicles over block pavements may be, to a great extent, avoided in parts of our towns which are devoted to residence. Moreover, the aggregate cost for the construction and maintenance of ways may by this means be somewhat reduced.

An incidental mention has already been made of the possible advantage arising from the use of large stone blocks on city pavements. It may be said that constructions of this same general nature are common in many parts of the Old World, and that, except for the slippery character of the surface in wet weather and in times of frost, there seems no objection to the method. In the condition of our modern art of stone-cutting by machinery it would be quite possible to maintain the surface of such large paving-blocks in a suitably roughened condition to insure a sufficient hold for the horses' feet by the use of machines which would from time to time channel the sur-

14

face in some proper manner. Such machines also could be used to plane down any irregularities of the surface which would arise from the diverse rates of wear of different parts of the stone.

The greatest difficulty which has been encountered in maintaining the highways of great cities arises from the use of the ground beneath them for a variety of purposes—sewerage, water-supply, gas, etc. There is but one method in which the exceeding evils which this entails can be overcome, and that is by having some form of large conduit below the street which will afford a place for all the subterranean structures which need traverse the way.

CHAPTER XII

ON EDUCATION IN THE SCIENCE AND ART OF ROAD-BUILDING

Conditions of a training in highway engineering. Methods of instruction. Opportunities for employment in highway engineering. Relations of State boards to highway engineers. Special methods of instruction

THE importance of extending knowledge concerning road-building in this country has often been incidentally mentioned in the preceding pages. This matter is of such importance that it deserves the especial consideration which will now be given to it. Let us first note the fact that in America there is no traditional knowledge of the subject such as has been gathered and transmitted in the Old World from centuries of experience and the teaching of able road-masters. Like traditions, as regards agriculture and the mechanic arts, were brought to America by the early settlers, have been deliberately imported by bringing skilled workmen of the several crafts, or have been locally developed in a way to meet our peculiar needs; but the art of road-making and road-keeping has, by reason of the newness of the country, together with the need of capital in the costly work of subjugating the land, remained undeveloped. In our time the problem refers to the means by which we may provide men with the training necessary to do the tasks of the road engineer

in an efficient manner. We need not be concerned about
the training of the foreman or the humbler laborers. It
is the province of the engineer, one which he easily fills,
to teach ordinarily intelligent workmen how to carry out
the plans he may form.

CONDITIONS OF A TRAINING IN HIGHWAY ENGINEERING

One of the principal difficulties in arranging for the
education of highway engineers arises from an entire mis-
conception as to the extent to which they need training.
Roads appear, to those who do not know either their importance or the complicated nature of the problems involved in their structure and maintenance, to be such very
simple things that, like other states of dust, such as the
soil itself, there appears to be needed only a little ordinary
practice to fit any dabster to deal with them. In fact, our
highways require of the engineers who are to deal with
them in an efficient way a wider range of knowledge than
is demanded of any other branch of engineering labor.
A thoroughly well-trained highway engineer must be a
good topographer, and a competent geologist as regards
many parts of that science, especially petrography and the
structure and history of the surface of the ground. He
should know something of climatology and chemistry. To
these acquirements he should add the resources of what
is commonly called civil engineering, as well as so much of
mechanical engineering as may fit him to deal with the
machinery which is used in his art. To fit him for duties
which pertain to so many branches of science and tech-
nology calls for a body of practical experience superadded
to a sound and extensive general training. He must, in
a word, be well educated before he can be deemed prepared

to deal with the apparently commonplace business of road-making. It is clear, in a word, that men are not to be fitted for this class of duties by mere practice. The task demands well-organized schooling.

It is a very characteristic error of our modern method of educating men for engineering work that we proceed to divide them into separate professional groups, making their training appear to relate to distinct accomplishments. We thus have the courses of study in our technical schools arranged with reference to degrees in civil, mechanical, electrical, hydraulic, and other engineering work, the system resting on the principle, or rather on the assumption, that quite diverse educational schemes are required to fit persons for these several employments. Something of the same motive is to be noted in other fields of education where men are to be trained for employments, but nowhere else is the evil of over-specialization of instruction so evident as in that of engineering. A true view of the matter, one to be attained with a little consideration, will show that, while the work of this profession is many-sided, the preparation for it demands of all students essentially the same plan of study.

The characteristic feature of the profession of engineering is that it has to do with the scientific application of the energies and resistances of the physical world to the needs of man. Thus defined, it stands apart from other branches of human activity in a fairly clear way. Other occupations, such as those of the surgeon, have a certain kinship with it, but they are distinguished by the fact that they relate to the organic or living side of the natural world. The common feature of all engineering work being the need of fitting the student for applying the arts which deal with the grosser and more massive side of the world's

14*

affairs, it is proper that the training which is to fit men for this field of duty should, whatever the special application of the work is to be, have a like foundation. All of them need to gain a clear sense of the properties of matter and the modes of action of energy, as well as the mathematical knowledge by which computations relating to these features may be made.

METHODS OF INSTRUCTION

For the reasons given above it is best that the training to be given to engineers who are to devote themselves mainly to highway work should in its essentials be the same as that which fits the other men who are to enter this large profession. They should, in a word, be well-trained civil engineers in the largest sense of that term. To this general study there should be added certain special teaching designed to meet the needs which the road-builder is sure to encounter. A good working knowledge as to the properties of rocks and of the nature of the surface deposits of the earth, valuable to all men of the engineering profession, is indispensable to those who are to deal in an intimate and effective way with earth and stones. The chemistry and physics needed by all engineers will suffice for those who are to have to do with this branch of the profession, but some little learning in climatology, especially that which pertains to rainfall and to the effects of climate on the surface of the earth, would be advantageous.

As regards the details of the instruction in road-making which should be given to those engineering students who may intend to make a specialty of highway work, I cannot give a clear idea so well in any other way as by stating the method which is followed in the Lawrence Scientific School of Harvard University. In that institution the in-

MERLINGEN ROAD, SWITZERLAND.

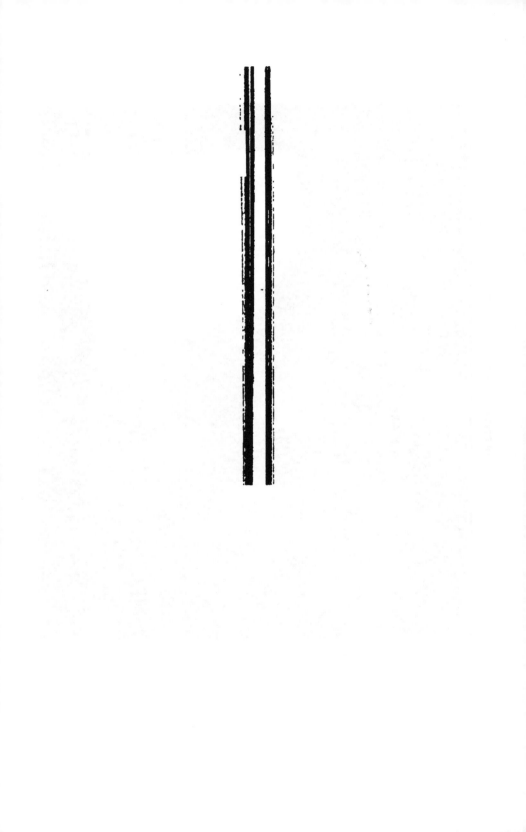

struction in this branch is given in the fourth or last year of the course for civil engineers. The instruction is given by Mr. W. E. McClintock, the engineer member of the Massachusetts Highway Commission, long and well known as an expert in road-building. The details of the system have been wrought out in connection with the experience gained as to the needs of engineers employed in that State. The teaching includes a course of lectures, illustrated by models and lantern slides, in which are considered the history of road-building and the various modes of doing such work. In the engineering laboratories of the school there is one devoted to experiments on the properties of road-building materials and the methods of testing them. In this workshop there has been carried on for several years a series of tests for the information of the Massachusetts Highway Commission. These experiments have led to the invention of methods and of apparatus which have brought about a considerable and important extension of the formerly existing means for exploring the properties of road-building materials. This apparatus and the practical work which it is doing serve for the instruction of students. During term time, as well as in the long vacation, the men have abundant opportunities of visiting roads which are under the process of construction and repair. The neighborhood of Cambridge, within a radius of five miles from the school, affords in its variety of roads, as regards their natural conditions and their treatment, opportunities for practical inquiry which are exceedingly good. There the students may see all grades of roads, from the various kinds of pavements used in a great city to the ordinary earth ways of very rural districts. They may observe the effects of good and bad systems of construction and maintenance. It may be said, in a word, that the opportunities for obtaining a clear and extended view of road problems are probably unequaled by any other field in this country.

Although the conditions for the development of instruction in the art of road-making are at present better at Harvard than they are at any other school in this country, and the project for such work has been carried out there in a more complete manner than elsewhere, there is no reason why in every other regard, except that which concerns the practical instruction based on the processes of road-building under varied conditions, like teaching should not be done in every large engineering school in this country. The cost of a well-fitted laboratory need not exceed ten thousand dollars. If, indeed, it be established in connection with those in petrography and mechanical engineering, it may be very much less than that sum. If the establishment is maintained in connection with State work in testing road-building materials, there will be an assurance of continued practical experiments which is necessary to give the highest value to illustrations such as a workshop may afford to students.

OPPORTUNITIES FOR EMPLOYMENT IN HIGHWAY ENGINEERING

It may well be asked, What is the chance of a young man finding employment, provided he prepares himself in a thoroughgoing way for the duties of a road engineer? In answer to this question, it may be said that at this moment the profession is about in the condition in which electrical engineering was twenty years ago. The attention of the public has been curiously awakened to the need of bettered ways. As yet the importance of having such ways built under the charge of competent experts has not become evident to the people, as it soon must.   Yet there is already a growing demand for persons of training in the highway art.   The Massachusetts Highway Commission last year employed about fifty men in the field, the greater number as resident engineers in charge of constructions. The extension of State work in other parts of the country appears assured.   It promises, indeed, within ten years to create a demand for the services of competent road surveyors much greater than that which is now made by any other one branch of the engineering profession.   Thus there is enough promise to justify the young engineer in devoting a portion of his study to those branches which may serve to fit him for highway work, though the opportunity does not justify him in accepting a training which would make him fit for this employment alone.   In fact, no chance whatever will warrant a student in limiting himself to one narrow field of what should be a broad and enlarging profession.

The employment of trained engineers in highway work promises to develop a field of occupation for men in this profession which will give its members something like the same station in our rural societies that is now held by physicians and lawyers, with quite as close a relation to their general life.   Hitherto there has not been enough business in the way of engineering in the districts of this country outside of the great towns to afford a livelihood to engineers.   We may now hope that well-trained men may find a substantial basis for support in the care of highways, and that they may incidentally take charge of a wide range of work in other branches of their profession which have hitherto been left without due attention.   Such matters as household and town drainage and water-supply would receive better consideration were engineers so distributed that they might enter in an intimate way into the councils of the people.

RELATIONS OF STATE BOARDS TO HIGHWAY ENGINEERS

The highway administration which may be established through State boards having such work in charge can do much to affirm the position of the rural engineers by arranging the work of supervision, and perhaps of construction as well, in such a manner that it will be in the hands of resident surveyors rather than in those of men who are merely detailed for special tasks. There is an advantage in having such tasks in charge of persons who dwell in the region in which the work lies, for the reason that they may come to have that detailed knowledge concerning the under-earth conditions on which sound practice in road-building so intimately depends. At present, however, there is no class of local engineers on whom the officers of these boards can rely. It will only be in a gradual manner that such an arrangement can be brought about.

SPECIAL METHODS OF INSTRUCTION

While awaiting the development of a body of well-trained engineers to whom may be committed the care of our highways, it is worth while to do what is possible to improve the condition of the usually untrained, but generally intelligent and devoted, men who have charge of our roads. The members of this class of superintendents of highways are necessarily more ignorant of their business than any other body of men who are charged with important public duties. From the nature of their conditions this could not be otherwise than it is.   With rare exceptions, they have never seen a road built which could in any way fitly serve to guide them in their own undertakings. Now and then one who journeys much will find a set of roads built by some born master who has worked out for himself the problems of drainage or the use of gravel and other materials which the roadside may afford; but such are very rare exceptions. The ordinary road-master of this country is not inventive, or, if so, his devices are of a vicious sort. For all their shortcomings, they are as a rule faithful public servants, very anxious to advance in their art.   Under these conditions we may hope to effect good results in the manner set forth below.

The superintendents of roads in each conveniently large district should be brought together once each year for instruction.   This could be given by means of lectures, and perhaps by practical illustrations showing the methods of road-building with the use of the standard means of construction.   The meetings might well be held at places where roads were in process of building or repair.   A provision in the laws regulating the Massachusetts commission requires that board once each year to hold an advertised meeting in the shire town of each county, at which meeting road-masters and others may seek information concerning the making and care of highways. Experience shows that these meetings are profitable to the men who resort to them for information.   The results make it seem likely that a more extensive and systematic plan of instruction might be as helpful to road-masters as the like methods of teachers' institutes are helpful in our schools. It may be that a suitable method of doing this work would be through the agricultural societies of this country.   In many of the States these societies are strong enough to afford the cost of this undertaking. At the moment there are too few persons acquainted with the work of road-building to provide instructors for such work, but in a few years we may hope to have experts at command for this duty.

In Massachusetts the highway masters have an association which has regular meetings, at which there are opportunities for discussing methods and results of their labors. In this, as in other employments, such meetings tend to develop a professional spirit, which always leads to better work on the part of those who feel it. There is no doubt that much good will be accomplished by founding such associations wherever the road-masters can be induced to interest themselves in the matter.

Although the literature concerning road-making which is fitted to the needs of this country is not large in amount, there are certain works which deserve a place in all rural libraries, where they may serve the needs of those who are interested in the matter. For the convenience of those who may desire to possess such works a list of the more important of those in the English language is given in an appendix. It should be said that until within a few years there have been few important contributions to this branch of economic science brought forth in this country, and that the French and German works, though numerous and interesting to the specialist, are generally of little value to our people, for the reason that they deal with conditions of labor, administration, and to a great extent of climate, which are quite other than our own. Even the English works, though they are perhaps the best for American use, do not completely meet the needs of this country.

CHAPTER XIII

SUMMARY AND CONCLUSION

Existing conditions of transportation in this country. Reasons for failure. Methods of amendment. Methods of Massachusetts Highway Commission. Effect of methods of local government on improvement of roads. Share of the federal government. Possible effect of inventions

THE treatment of the problems concerning American roads which has been presented in the previous chapters of this book has, because of the nature of the subject, been somewhat discursive. It therefore seems desirable to sum up the various considerations which have been deduced, and to draw from them the conclusions which may promise to be of value.

Those who would understand the state and importance of the American highway question should consider how intimately the economic success and the social development of communities depend upon the ease with which the people maintain the commerce within their society and with the outer world. So absolute in this day is this dependence on ease of traffic that each distinct improvement or hindrance of it is at once reflected in many and varied ways; not alone in those of trade, but also in moral and intellectual development.

EXISTING CONDITIONS OF TRANSPORTATION IN THIS COUNTRY

So far as those methods of transportation which relate to the distant intercourse of our people are concerned, this country compares, as a whole, very favorably with any equally extensive area. Judged on the basis of population, however, it is still somewhat less well provided with railways than are several European states; yet the rate of growth of these means of distant carriage is such as to insure within one or two decades that the iron ways which serve our people will constitute as perfect a system as exists in any land.

While the railway and steamboat system of the United States has been developed with great rapidity, and in a way that reflects much credit on the skill and energy of our business men, the ancient and more necessary means of transportation afforded by the ordinary wagon roads have remained in a state of shameful neglect. Thus, while during this half-century the usefulness of routes for national and international intercourse has been vastly increased as regards speed, cheapness, and safety, the highways which are tributary to them are, as a whole, in a poorer state than they were fifty years ago.

REASONS FOR FAILURE

It is not easy to account for the failure of the American people to attain success in the management of their highways. It is evident, however, that we cannot charge the lack of accomplishment to a deficiency of public spirit or an incapacity to deal with public affairs. The most reasonable explanation of the situation is apparently to be found mainly in the exceeding ignorance which prevails as to the nature and value of good roads, and in lesser measure in the obvious defects in our system of local government. As to the first and greatest of these hindrances much has been said in the preceding chapters of this book. The effort has been made to show that this country was settled and took its shape before the revival of road-making in Europe came about after the dark ages, during which the governmental conditions of the Old World made against the free intercourse of peoples. Of all the arts which relate to the welfare of societies, that of highway construction was the last to be restored to the state in which it existed at the downfall of the Roman empire. This restoration was slowly effected, the work being mainly done by the French. It was not in any considerable measure accomplished until the Revolutionary War had separated the American colonies from European influence. Thus left to themselves, our people, to a great extent occupied with other immediate interests, have never earnestly attended to the problem of highways. Even the movement for the betterment of roads which is now observable in this country has not as yet affected the masses of the people who are concerned with the matter ; it is still limited, as far as action goes, to a few States lying in the north-eastern part of the Union.

METHODS OF AMENDMENT

As regards the amendment of our system, it is evident that the first important step which needs to be taken is one that will educate the people as to the means whereby their roads may be improved. To attain this end they must be convinced that the task of constructing good and enduring ways is much more serious than they believe it to be, and that its proper accomplishment calls for a systematic training of the men who are to do the work. They must be brought to see that such undertakings demand the services of educated men quite as much as any other engineering work, such as bridge-building or railway construction.   Unless this end is attained, we may expect to see, as the result of the present extending interest in roads, some increase in our present annual expenditure on rural ways, followed by a speedy decline of the hope that larger taxes for highways would be profitable.   This danger of disappointment, due to injudicious expenditure, gravely threatens to bring the present movement for bettered ways to an unhappy end.

The data for determining the annual expenditure of money on the roads of this country have never been collected with sufficient accuracy to permit a close reckoning as to this element of our national taxation.   Enough is known, however, to warrant the assertion that the total, excluding the cost of city streets, exceeds fifty million dollars, and may amount to seventy-five.   Of this vast sum the greater part is wasted in ill-contrived and temporary repairs needed to keep the ways in a condition that will permit the commerce which is required to maintain our civilization.   In the present state of our taxation, with the burden of the very complicated and costly system of government, and a pension debt which may be reckoned at some thousand million dollars, resting upon the people, it is too much to expect that they will speedily effect the task of reconstructing their highways.   This nation, as a whole, is in the position occupied by a farmer who has broad and fertile acres which have been recklessly mortgaged up to the measure of their net production, yet who needs to renew his barns and tools. It is useless to tell him that he should have better appliances, or even to show him that he could make more money with such a provision. His reasonable answer is that such improvements must wait until he has lessened the load of his debts.

Assuming that the people of this country, except in the older and richer parts, cannot at present well afford to shoulder the debt which it would be necessary to incur in order to obtain a substantial betterment of their highways, the question arises as to how far, on something like the present scale of expenditure on roads, it may be possible to make a reasonable advance on the existing conditions. It is evident that this end can be attained, if at all, only by diverting to permanent constructions a portion of the money which is now paid for merely temporary improvements. A close study of the conditions of a number of New England towns and of certain districts in the central and southern portions of the Union has shown me that it is possible so to change the method of highway work as to attain the end in view. The method of doing this may be briefly set forth as follows:

Within any governmental unit of highway management, town, precinct, parish, or county, the roads should be mapped in any convenient manner (for this purpose the charting need not be very accurate), so that they may be classified according to their economic importance. It will in most districts be found that somewhere about one eighth of the total mileage of way belongs in the group of main routes which concern the community as a whole; about four eighths fall into the middle group, which serve the interests of large neighborhoods, while the remainder are mere branches of the more important highways, designed for the accommodation of a few farms. With

15

such a classification graphically presented by means of a map, a selection should be made of those parts of the first group which are of critical importance. In most districts there are some very bad pieces of main road which from their difficulty have placed a tax on all the commerce of the community since it was founded. By annually devoting a certain share of the money which is applied to highways to the effective improvement of these stumbling-blocks to travel, it is possible in a decade greatly to elevate the condition of the roads in the neighborhood without either materially increasing the taxes or seriously neglecting the repairs of the less important ways.

When the financial condition of a community makes it fit that road construction should be undertaken more rapidly than would be found possible by the method above indicated, it will be well to raise the money for the proposed improvement either from a special tax or by a loan to be devoted to this particular use. With either of these methods the cost should be distributed over a considerable term of years. As already noted in the preceding chapters, the conditions determining road-making vary greatly in different parts of this country. The methods of work have to be adapted to these diverse circumstances if due economy is to be secured. Very slight differences in the state of the undersoil, or in the nature of the stone used in hardening the surface, have to be regarded. It is most desirable that experience as regards these and other critically important matters should be acquired at the least possible cost from mistaken efforts. To attain this end it is necessary to avoid haste in improving the roads of any country; time needs to be allowed for the results of the first inevitably experimental essays to be shown.

METHODS OF MASSACHUSETTS HIGHWAY COMMISSION

An application of the tentative method of highway construction was made on a tolerably large scale by the Massachusetts commission in beginning the construction of State roads in that commonwealth. Although there seemed to be a general demand that a limited number of main roads should be undertaken and carried straight across the State, the adoption of such a policy would have led to building these ways without sufficient knowledge of the conditions which were to be encountered, and would have involved many blunders. The commission chose—wisely, as the issue showed—to begin by building about seventy pieces of roadway, having an average length of a mile each. These bits were so placed in different parts of the State that, while they would in time fit into a system of State roads, they served as tests as to the suitability of materials and methods to meet the peculiar needs of each locality.

One of the most important results of the essays in road-building above noted is the evidence that the hardened part of the way, as it is usually built in this country as well as in most parts of Europe, is much wider than it is necessary to have it. While it is clearly desirable to have the location of a public road in most cases wide enough to admit of sidewalks and, it may be, of an electric tramway, the present needs can often be met in a very satisfactory manner by ways not more than from ten to twelve feet in width, with graveled or grassed shoulders having a width of four feet on each side of the macadamized strip. It therefore seems desirable, where, as is generally the case, there is need of extreme economy in the construction, to build very much narrower hardened ways than our people are accustomed to accept.

It has recently been urged by several writers that the improvement of roads should be undertaken by local societies, which, organized for this specific purpose, should be allowed by the local authorities to dispose of a part of the public money which is to be applied to highway work. It may be held, however, that, while such associations may well be agents for stimulating the officials who are responsible for building and keeping public ways, they should not have to do with the actual construction. It is not to be expected that any self-constituted society can have the continuous life, the representative quality, and the other elements of power which belong to the body politic. In that body alone, changed if needs be as regards its mechanism, and stimulated, as is usually required, by the urgence of good citizens, we must always put our trust for the betterment of public affairs. In such unessential matters as the decoration of roadways by plantations, etc., village improvement societies, or those of larger scope, may well act with profit.

EFFECT OF METHODS OF LOCAL GOVERNMENT ON IMPROVEMENT OF ROADS

As at present constituted, the government of our rural communities outside of New England is not well fitted to meet the needs of our people; this is shown not only in the matter of highway work, but in all the functions which relate to the care of local interests. In New England the town system, that in which each community governs itself by a local parliament, the orders of which are carried out by the selectmen, affords a means whereby the judgments and criticism of the people can be effectively applied to local affairs. It is otherwise with the States which lie to the west and south of the Hudson River. In those communities the local government is generally effected by the county system, with special but imperfect arrangements for the care of certain interests through local boards for villages and school districts. Rarely is there any trace of the free debate and immediate reference to the people of all questions which the town-meeting provides for. Our usual American county system is in effect a very clumsy adaptation of an ancient English system of administration which originated in conditions totally different from our own, and which imperatively needs to be reformed before it will be possible to bring the machinery of our rural government into satisfactory shape.

It is, perhaps, not too much to hope that the value of the New England town government may, by the demand which the movement toward bettered ways is making for a more effective method of local government, induce our people generally to adopt the citizens' parliament. It may be noted in passing that the changes in the machinery of our States which would be required in order to institute such a plan of control are not great, if the intent were limited to the matters connected with roads. It would probably be accomplished in most of the States without any change in the organic law. To attain the end the counties should be organized in road districts, each acting through a general assembly of the voters, who should elect their selectmen. To this executive board should be committed the expenditure of the money allotted from the general tax-list of the county, or, what would be better, levied by the district on the property within its

15*

limits. If a method of control of highway business could be established on the lines which have above been indicated, there would be reason to expect that in time it would be extended to other affairs. In this manner it may be possible to implant parliamentary local government in those parts of the country where it does not exist, and where there is little chance for its introduction in its complete form.

Where the States of this Union undertake to build the main highways, it is most desirable, for reasons already set forth, that the work should be done, not by private contractors, but by representatives of the people; for where it is placed in the hands of such selectmen the task will be performed under conditions which favor the dissemination of knowledge concerning the art, which makes the work highly educative. It has been most interesting to see how rapidly the State work in Massachusetts has developed a body of men who are able to take charge of road-building in at least one method of construction. When the commonwealth began to build roads there were no people, outside of a few engineers' offices, who knew anything about the business. It is safe to reckon that there are at the present time not less than five hundred who have taken a sufficiently intelligent share in the labor to be tolerably well trained in the routine part of the business.

While the conduct of work by the State, provided the tasks are done under the direction of skilled engineers, will serve to arouse popular interest in the improvement of roads and disseminate knowledge of the art, thereby laying the foundations of a better system, it must not be supposed that we can trust to it alone. To insure the needed learning it will be necessary for our schools where engineering is taught to train experts in highway construction who should be employed in determining the ways and means of each improvement. Not only is it necessary to have sound expert advice in the choice and methods of use of the materials, but the location of new ways and the modification of the position of those now in use call for a peculiar skill.

Where a State deems it inadvisable to pay all the costs of the main highways which it may undertake to build, as is likely to be the case with the greater number of our commonwealths, it will still be well for the work of construction to be in the hands of a central board. It has been proposed that the contribution of the State treasury should be in some way handed over to local authorities to expend. While this method may be practicable after a knowledge of road-building is sufficiently diffused, it would at present be very ill fitted to the needs. The most important result arising from well-managed central systems is that the ignorance of our people in such matters may be broken down, and that they may come to know of the existence of a science and art of highway-building.

SHARE OF THE FEDERAL GOVERNMENT

As the road problem is a matter of national importance, it is worth while to consider what share, if any, the federal government may properly have in the work. If wagon roads were to any considerable extent the paths of interstate commerce, or if they were possibly to be reckoned on for marching troops, there would be good reason to expect that certain, perhaps many, main ways would be built and maintained in whole or in part at the cost of the nation. The changes in the methods of transportation which have been effected during the last half-century have made ordinary roads mere adjuncts to the railways. It rarely occurs that merchandise is wagoned for more than thirty miles. The average distance of such transportation for agricultural products is probably not more than about five miles. There is no systematic communication between the States by ordinary roads. In no one case is the intercourse between the capitals of two of these units in the federal Union kept up by the use of highways. In case of war, whether it were internecine or with a foreign power, all distant transportation would commonly be accomplished by the railways or rivers, as it was, indeed, during the Rebellion. For short distances armies, for the reason that they are not troubled by economic considerations, are peculiarly independent of the conditions of the roadways. The ills they present weigh alike on both the combatants. The only point at which the federal system touches the highway problem concerns the carriage of the mails.

An inspection of the mail-route maps will show that the greater number of the main roads in the States outside of New England are used for transporting the mail. If this use is taken as a justification of action from Washington, it might be held that from there should come the means for the improvement of these ways. It is evident, however, that, as the mails with us carry very little merchandise, being in this regard different from those in several European countries, the amount of transportation, measured by weight, is insignificant. In almost all cases the mail-bags are carried in some public conveyance, the service being a mere adjunct to other business. All that the central government would gain by bettered roads would be a somewhat speedier transit of the property which it undertakes to deliver. This is hardly enough of an advantage to warrant the institution of a national work, which would prove extremely costly, after the manner of all federal constructions, and which would require an army of public servants for its administration. However effected, the people have in the end to pay for what they get from the powers above them. The price is the dearer the further these powers are removed from local inspection. It therefore does not seem advisable to seek federal aid in the work of constructing our ways.

It has been suggested that the United States engineers might well be employed in time of peace—which we may well expect to be for the greater part, if not all, of their lives—in supervising the construction and maintenance of highways. The objection to this proposition is that these selected men are trained for a difficult and peculiar profession, one which demands so much and so varied knowledge that it would be impossible for them to add the large store of information which the highway expert has to acquire. Moreover, the army engineer is very properly trained to accomplish the tasks which are intrusted to him with very little regard to the cost of the work, expense in matters of national defense being a matter of very little importance. The result of these conditions of education is that, while the engineering corps of our army is a remarkably able and well-trained body, it is not one to which we may look for the men who are to take charge of works which need to be done with the utmost economy.

The only way in which the federal government can fairly give aid in the improvement of roads is by the dissemination of information concerning the form of the surface of the country, and the nature and distribution of the various materials which may be used in their construction. This is in process of doing by the United States Geological Survey, which by its topographical maps and the reports based thereon is rapidly making a provision of knowledge which will be of distinct value to the roadbuilder. To this printed matter there might well be added reports on the road-building stones of each important district. It may also be well to provide a laboratory in which such materials could be tested in the manner noted in a previous chapter, so that these determinations might have a uniform value for all parts of the country.

Among the many suggestions concerning the changes in the conditions of our highways which are likely soon to be brought about by the progress of invention, there are two which, on account of their evident importance, deserve some mention. The first of these relates to the probable effect of the application of some form of mechanical energy to the propulsion of carriages; the other to the adoption of steel tramways for the use of ordinary freight-wagons. Should it come about that power-carriages displaced those drawn by animals, a large part of the difficulties which the roadmaster has to encounter would be obviated. The costs of repairs, so far as is due to the wearing action of the shod feet, would of course disappear, and at the same time the need of reducing the grades to a low angle would probably be less great than it is at present.

As for the suggestion that grooved steel rails might advantageously be used in such a way as to afford a favorable path for heavily laden wagons, it may be said that a certain amount of experiment has been made to this end, though it has never had a thoroughgoing trial.   The objections to the method are that it would require all the wagons going in one direction to maintain the same rate of movement, and that the structure would be exceedingly costly.   Such a double track on the average could not well be built for less than twenty thousand dollars per mile; a sum to be reckoned in addition to that required for the construction of the ordinary way designed for the lighter vehicles.   It is evident that a better solution of the problem of transporting freight is likely to be found in the use for that purpose of the electric tramways which, as before noted, are likely to parallel every important highroad.   If this method is used at all, it must be under very exceptional conditions.

The matter presented to the reader in the preceding pages, though it by no means includes all the considerations which relate to the problems of road-making in this country, is perhaps sufficient to show something of the importance of the subject.   Owing to certain peculiar accidents of our national history, ours is, of all civilized lands, the most belated in this element of material and social development.   Our situation in this regard is not only, in a way, disgraceful, but it is in high measure unfortunate.   We are undertaking to maintain a social and economic order which shall give the individual citizen a measure of opportunities such as no other country has ever proposed to afford to its people; yet we lack the most important means for the accomplishment of the purpose, in that our ways of communication are not in condition to serve our needs.   It is evident that something like an arrest in the development of this country marks the closing years of the century.   This situation is doubtless due to many causes, but it is not unreasonable to reckon among them the excessive cost of all kinds of work, which is due to the great transportation tax which our people have to bear, a burden which, so far as our agricultural work is concerned, may fairly be reckoned as greater than all the other imposts they endure. It is clearly the first duty of statesmanship to lighten this ancient and grievous burden, and in the effort for the relief every good citizen may fairly be expected to take a willing share.

APPENDICES

APPENDIX A

ACTS OF THE LEGISLATURE OF MASSACHUSETTS RELATING TO STATE HIGHWAYS

As the commonwealth of Massachusetts has been for several years engaged in a systematic and extensive way in improving its main roads, it seems desirable to assemble all the important acts which relate to this work. These acts are here for the first time published together.

ACTS OF 1892, CHAPTER 338

AN ACT TO ESTABLISH A COMMISSION TO IMPROVE THE HIGHWAYS OF THIS COMMONWEALTH

Be it enacted, etc., as follows :

SECTION 1. The governor, with the advice and consent of the council, shall, within thirty days from the passage of this act, appoint three persons, one of whom shall be a civil engineer, whose term of office shall expire on the first Wednesday of February in the year eighteen hundred and ninety-three, to consider what legislation is necessary for the better construction and maintenance of the highways in this commonwealth.

SECTION 2. The said commission shall forthwith proceed to investigate and consider the best and most practicable method of construction and maintenance of highways, and the estimated cost of the various methods and systems ; the establishment of State or county highways, with recommendations as to their construction and maintenance ; routes and the approximate cost ; also the geological formation so far as it relates to the material suitable and proper for road-building. Said commission shall prepare suitable maps and plans, on which shall be clearly drawn the various routes they recommend.

SECTION 3. Said commission may establish rules and regulations for the conduct of its business, and shall be provided with suitable quarters by the sergeant-at-arms in the state-house or elsewhere. They may employ experts and all necessary clerical and other assistants, and may incur such reasonable expenses, including traveling expenses, as may be authorized by the governor and council. Before incurring any expenses they shall from time to time estimate the amount required, and shall submit the same to the governor and council for their approval; and no expense shall be incurred by the commission beyond the amount so estimated and approved. Said commission shall receive such compensation as the governor and council may decide, provided the whole amount expended under the provisions of this act shall not exceed ten thousand dollars.

SECTION 4. The county commissioners, boards of selectmen and aldermen, and other officers having authority over public ways, roads, and bridges throughout the commonwealth, shall at reasonable times, on request, furnish the commissioners any information required by them concerning the public ways, roads, or bridges within their jurisdiction. The commissioners may furnish blank forms for returns to be made to them by such officers, and may make changes in and additions to such forms.

SECTION 5. The said commission shall report fully with plans and estimates and their recommendations to the legislature on or before the first Wednesday of February in the year eighteen hundred and ninety-three, and shall append to its report a draft of a bill intended to accomplish the recommendations of the commission.

SECTION 6. Any vacancy in the commission may be filled by the governor, with the advice and consent of the council.

SECTION 7. This act shall take effect upon its passage.

[*Approved June* 2, 1892.]

RESOLVES OF 1893, CHAPTER 45

RESOLVE PROVIDING FOR PRINTING THE REPORT OF THE
COMMISSION TO IMPROVE THE HIGHWAYS
OF THIS COMMONWEALTH

Resolved, That there be electrotyped, printed, and bound in cloth five thousand copies of the report of the commission appointed to consider what legislation is necessary for the better construction and maintenance of the highways in this commonwealth, with appendices marked A, B, C, D, E, F, G, H, I, J, and K, containing the statistics in tabulated form prepared under the direction of said commission, to be distributed as follows: to each member of the General Court ten copies; to each member of the executive departments, the clerks and assistant clerks of the two branches of the General Court, and each reporter assigned a seat in either branch, one copy; to the State Library twenty-five copies; to the Massachusetts Historical Society and the New England Historic Genealogical Society five copies each; to each free public library in the commonwealth which is open to the use of the city or town where it is situated one copy; all of which shall be distributed under the direction of the secretary of the commonwealth; the balance shall be placed in the office of the secretary of the commonwealth for public distribution.

[*Approved March* 29, 1893.]

ACTS OF 1893, CHAPTER 476

AN ACT TO PROVIDE FOR THE APPOINTMENT OF A HIGHWAY COMMISSION TO IMPROVE THE PUBLIC ROADS, AND TO DEFINE ITS POWERS AND DUTIES

Be it enacted, etc., as follows:

SECTION 1. The governor, with the advice and consent of the council, shall, within thirty days after the passage of this act, appoint three competent persons to serve as the Massachusetts Highway Commission. Their terms of office shall be so arranged and designated at the time of their appointment that the term of one member shall expire in three years, one in two years, and one in one year. The full term of office thereafter shall be for three years, and all vacancies occurring shall be filled by the governor, with the advice and consent of the council. The members of said board may be removed by the governor, with the advice and consent of the council, for such cause as he shall deem sufficient and shall express in the order of removal. They shall each receive in full compensation for their services an annual salary of two thousand dollars, payable in equal monthly instalments, and also their traveling expenses. They may expend annually for clerk hire, engineers, and for defraying expenses incidental to and necessary for the performance of their

16

duties, exclusive of office rent, the sum of two thousand dollars.
They shall be provided with an office in the state-house, or some
other suitable place in the city of Boston, in which the records of
their office shall be kept.   They may establish rules and regulations
for the conduct of business and for carrying out the provisions of
this act.

SECTION 2. They shall from time to time compile statistics relating
to the public roads of cities, towns, and counties, and make such
investigations relating thereto as they shall deem expedient.   They
may be consulted at all reasonable times without charge by officers
of counties, cities, or towns having the care of and authority over
public roads, and shall without charge advise them relative to the
construction, repair, alteration, or maintenance of the same; but
advice given by them to any such officers shall not impair the legal
duties and obligations of any county, city, or town.   They shall prepare a map or maps of the commonwealth on which shall be shown
county, city, and town boundaries, and also the public roads, particularly the State highways, giving when practicable the names of
the same.   They shall collect and collate information concerning the
geological formation of this commonwealth so far as it relates to the
material suitable and proper for road-building, and shall, so far as
practicable, designate on said map or maps the location of such material.   Such map or maps shall at all reasonable times be open for
the inspection of officers of counties, cities, and towns having the
care of and authority over public roads.   They shall each year hold
at least one public meeting in each county for the open discussion
of questions relating to the public roads, due notice of which shall
be given in the press or otherwise.

SECTION 3. They shall make an annual report to the legislature of
their doings and the expenditures of their office, together with such
statements, facts, and explanations bearing upon the construction
and maintenance of public roads, and such suggestions and recommendations as to the general policy of the commonwealth in respect
to the same, as may seem to them appropriate.   Their report shall
be transmitted to the secretary of the commonwealth on or before
the first Wednesday in January of each year, to be laid before the
legislature.   All maps, plans, and statistics collected and compiled
under their direction shall be preserved in their office.

SECTION 4. County commissioners and city and town officers having
the care of and authority over public roads and bridges throughout

the commonwealth shall, on request, furnish the commissioners any information required by them concerning the roads and bridges within their jurisdiction.

SECTION 5. For the purpose of carrying out the provisions of this act, said commission may expend such sums for necessary assistants, the procuring of necessary supplies, instruments, material, machinery, and other property, and for the construction and maintenance of State highways, as shall from time to time be appropriated by the legislature; and they shall in their annual report state what sums they deem necessary for the year commencing with the first day of March following.

SECTION 6. Whenever the county commissioners of a county adjudge that the common necessity and convenience require that the commonwealth acquire as a State highway a new or an existing road in that county, they may apply by petition in writing to the Massachusetts Highway Commission, stating the road they recommend, and setting forth a detailed description of said road by metes and bounds, together with a plan and profile of the same. Said commission shall consider such petition, and if they adjudge that it ought to be allowed, they shall in writing so notify said county commissioners. It shall then become the duty of said county commissioners to cause said road to be surveyed and laid out in the manner provided for the laying out and alteration of highways, the entire expense thereof to be borne and paid by said county. Said county commissioners shall preserve a copy of such petition, plans, and profiles, with their records, for public inspection. When said commission shall be satisfied that said county commissioners have properly surveyed and laid out said road, and set in place suitable monuments, and have furnished said commission with plans and profiles, on which shall be shown such monuments and established grades, in accordance with the rules and regulations of said commission, said commission may approve the same, and so notify in writing said county commissioners. Said commission shall then present a certified copy of said petition, on which their approval shall be indicated, together with their estimates for constructing said road and the estimated annual cost for maintaining the same, to the secretary of the commonwealth, who shall at once lay the same before the legislature if it is in session; otherwise on the second Wednesday of January following. If the legislature makes appropriation for constructing said road, said commission shall cause said road to be constructed in accordance with this act, and when completed and approved by them said road shall become a State highway and thereafter be maintained by the commonwealth under the supervision of said commission.

SECTION 7. Two or more cities or towns may petition the said commission, representing that in their opinion the common necessity and convenience require that the commonwealth should acquire as a State highway a new or an existing road leading from one city or town to another, which petition shall be accompanied by a detailed description of such road by metes and bounds, and also a plan and profile of the same. If said commission adjudge that the common necessity and convenience require such road to be laid out and acquired as a State highway, they shall cause a copy of said petition, on which shall be their finding, to be given to the county commissioners of the county in which said road or any portion of it lies. It shall then become the duty of the county commissioners, at the expense of the county, to cause said road to be surveyed and laid out, and to set in place suitable monuments, and to cause a detailed description by metes and bounds, plans and profiles, to be made, on which shall be shown said monuments and established grades, and to give the same to said commission; but said county commissioners shall have the right to change the line of said road, provided the termini are substantially the same. Said county commissioners shall preserve said petition and a copy of the plans and profiles, with their records, for public inspection. When said commission shall be satisfied that the county commissioners have properly surveyed and laid out said road and set in place suitable monuments, and have furnished them with plans and profiles on which shall be shown said monuments and established grades, in accordance with the rules and regulations of said commission, they shall then proceed in the same manner as provided in section six of this act; and when said road is completed and approved by said commission, it shall become a State highway, and thereafter be maintained by the commonwealth under the supervision of said commission.

SECTION 8. In all cases where a highway is to be constructed at the expense of the commonwealth as a State highway, all the grading necessary to make said highway of the established grade, and the construction of culverts and bridges, shall be paid for by the county or counties, respectively, in which said highway or any portion of it lies, and the work must be done to the satisfaction of said commission. No action by a person claiming damage for the taking of land or change of grade under the provisions of this act shall be commenced against a county until said commission has taken possession for the purpose of constructing such State highway.

SECTION 9. When appropriation has been made by the legislature for the construction of a State highway, said commission shall at once cause plans and specifications to be made, and estimate the cost of the construction of such State highway, and give to each city and town in which said road lies a certified copy of said plans and specifications, with a notice that said commission is ready for the construction of said road. Such city or town shall have the right, without advertisement, to contract with said commission for the construction of so much of such highway as lies within its limits, in accordance with the plans and specifications of the commission and under its supervision and subject to its approval, at a price agreed upon between said commission and said city or town; but such price agreed upon shall not exceed eighty-five per cent. of the original estimate of said commission. If such city or town shall within thirty days not elect to so contract, said commission may advertise in one or more papers published in the county where the road or portion of it is situated, and in one or more papers published in Boston, for bids for the construction of said highway, in accordance with the plans and specifications furnished by said commission and under their supervision and subject to their approval. Said commission shall have the right to reject any and all bids, and they shall require of the contractor a bond for at least ten thousand dollars for each mile of road, to indemnify such city or town in which such highway lies against damage while such road is being constructed, and the commonwealth shall not be liable for any damage occasioned thereby. Said commission shall make and sign all contracts in the name of the Massachusetts Highway Commission.

SECTION 10. For the maintenance of State highways said commission shall contract with the city or town in which such State highway lies, or a person, firm, or corporation, for the keeping in repair and maintaining of such highway, in accordance with the rules and regulations of said commission and subject to their supervision and approval, and such contracts may be made without previous advertisement.

SECTION 11. All contracts made by or with the Massachusetts Highway Commission under the provisions of this act shall be subject to the approval of the governor and council.

16*

SECTION 12. No length of possession or occupancy of land within the limit of any State highway by an owner or occupier of adjoining land shall create a right to such land in any adjoining owner or occupant or a person claiming under him; and any fences, buildings, sheds, or other obstructions encroaching upon such State highway shall, upon written notice by said commission, at once be removed by the owner or occupier of adjoining land, and if not so removed said commission may cause the same to be done and may remove the same upon the adjoining land of such owner or occupier.

SECTION 13. The commonwealth shall be liable for injuries to persons or property occurring through a defect or want of repair or of sufficient railing in or upon a State highway.

SECTION 14. Cities and towns shall have police jurisdiction over all State highways, and they shall at once notify in writing the State commission or its employees of any defect or want of repair in such highways. No State highway shall be dug up for laying or placing pipes, sewers, posts, wires, railways, or other purposes, and no tree shall be planted or removed or obstruction placed thereon, except by the written consent of the superintendent of streets or road commissioners of a city or town, approved by the Highway Commission, and then only in accordance with the rules and regulations of said commission; and in all cases the work shall be executed under the supervision and to the satisfaction of said commission, and the entire expense of replacing the highway in as good condition as before shall be paid by the parties to whom the consent was given or by whom the work was done; but a city or town shall have the right to dig up such State highway without such approval of the Highway Commission where immediate necessity demands it, but in all such cases such highway shall be at once replaced in as good condition as before and at the expense of the city or town. Said commission shall give suitable names to the State highways, and they shall have the right to change the name of any road that shall have become a part of a State highway. They shall cause to be erected at convenient points along State highways suitable guide-posts.

SECTION 15. The word "road" as used in this act includes every thoroughfare which the public has a right to use.

SECTION 16. This act shall take effect upon its passage.

[*Approved June* 10, 1893.]

ACTS OF 1894, CHAPTER 497

AN ACT RELATING TO STATE HIGHWAYS

Be it enacted, etc., as follows:

SECTION 1. Whenever the county commissioners of a county, or the mayor and aldermen of a city, or the selectmen of a town, adjudge that the public necessity and convenience require that the commonwealth take charge of a new or an existing road as a highway, in whole or in part, in that county, city, or town, they may apply by a petition in writing to the Massachusetts Highway Commission, stating the road they recommend, together with a plan and profile of the same.

SECTION 2. Said Highway Commission shall consider such petition and determine what the public necessity and convenience require in the premises, and if they deem that the highway should be laid out or be taken charge of by the commonwealth shall file a plan thereof in the office of the county commissioners of the county in which the petitioners reside, with the petition therefor and a certificate that they have laid out and taken charge of said highway, in accordance with said plan, and shall file a copy of the plan and location of the portion lying in each city or town in the office of the clerk of said city or town; and said highway shall, after the filing of said plans, be laid out as a highway and shall be constructed and kept in good repair and condition as a highway by said commission, at the expense of the commonwealth, and shall be known as a State road, and thereafter be maintained by the commonwealth under the supervision of said commission. And all openings and placing of structures in any such road shall be done in accordance with a permit from said commission.

SECTION 3. The damages sustained by any person whose property is taken for or is injured by the construction of any such highway shall be paid by the commonwealth, the same to be determined by said commission. And if said commission and the person sustaining the damages cannot agree thereon, he or they may have said damages determined by a jury in the county in which the land is situated, by filing a petition for such jury in the office of the clerk of the superior court for said county at any time before the expiration of one year from the completion of said highway, and thereupon said damages shall be determined by a jury at the bar of said court, in the same manner as damages for the taking of land for other highways in the county, city, or town are determined; and costs shall be taxed to the prevailing party on such petition, as in civil cases.

SECTION 4. Said commission shall, when about to construct any highway, give to each city and town in which said highway lies a certified copy of the plans and specifications for said highway, with a notice that said commission is ready for the construction of said road. Such city or town shall have the right, without advertisement, to contract with said commission for the construction of so much of such highway as lies within its limits, in accordance with the plans and specifications and under its supervision and subject to its approval, at a price agreed upon between said commission and said city or town. If said city or town shall not elect to so contract within thirty days, said commission shall advertise in two or more papers published in the county where the road or portion of it is situated, and in three or more daily papers published in Boston, for bids for the construction of said highway under their supervision and subject to their approval, in accordance with plans and specifications to be furnished by said commission. Such advertisement shall state the time and place for opening the proposals in answer to said advertisements, and reserve the right to reject any and all proposals. All such proposals shall be sealed and shall be kept by the board, and shall be open to public inspection after said proposals have been accepted or rejected. Said commission may reject any or all bids, or if a bid is satisfactory they shall, with the approval of the governor and council, make a contract in writing on behalf of the commonwealth for said construction, and shall require the contractor to give a bond for at least twenty-five per cent. of the contract price to indemnify any city or town in which such highway lies against damage while such road is being constructed, and the commonwealth shall not be liable for any damage occasioned thereby. All construction of State roads shall be fairly apportioned by said commission among the different counties, and not more than ten miles of State road shall be constructed in any one county in any one year on petition, as aforesaid, without the previous approval thereof in writing by the governor and council.

SECTION 5. One quarter of any money expended under the provisions of this act in any county for a highway, with interest on said quarter at the rate of three per cent. per annum, shall be repaid by said county to the commonwealth in such reasonable sums and at such times within six years thereafter as said commission, with the approval of the State auditor, shall determine, taking into consideration the financial condition of the county; and the treasurer and receiver-general shall apply all money so repaid to the appropriation to be expended by said commission. The county treasurer, with the approval of the county commissioners, may make such loans as they may see fit to meet this expenditure.

SECTION 6. Any city or town in which a State highway is situated shall be liable for injuries to persons traveling upon a State highway the same as upon other highways, but the amount actually recovered as damages for such injuries shall be repaid within one year thereafter to such city or town by the commonwealth. A city or town may make temporary necessary repairs of a State highway without the approval of said commission.

SECTION 7. Said commission shall keep all State roads reasonably out of brush, and shall cause suitable shade-trees to be set out along said highways when feasible, and shall renew the same when necessary, and may also establish and maintain watering-troughs at suitable places along said highways.

SECTION 8. For the purpose of meeting any expenses that may be incurred under the provisions of chapter four hundred and seventy-six of the acts of the year eighteen hundred and ninety-three, as hereby amended, including the salaries and expenses of the commission, the treasurer and receiver-general is hereby authorized, with the approval of the governor and council, to issue scrip or certificates of indebtedness to an amount not exceeding three hundred thousand dollars, for a term not exceeding thirty years. Said scrip or certificates of indebtedness shall be issued as registered bonds or with interest coupons attached, and shall bear interest not exceeding four per centum per annum, payable semi-annually on the first days of April and October in each year. Such scrip or certificates of indebtedness shall be designated on their face as the State Highway Loan, shall be countersigned by the governor, and shall be deemed a pledge of faith and credit of the commonwealth, and the principal and interest shall be paid at the times specified therein in gold coin of the United States or its equivalent; and said scrip or certificates of indebtedness shall be sold and disposed of at public auction, or in such other mode, and at such times and prices, and in such amounts, and at such rates of interest, not exceeding the rate above specified, as shall be deemed best. The treasurer and receiver-general shall, on issuing any of said scrip or certificates of indebtedness, establish a sinking-fund for the payment of said bonds, into which shall be paid any premiums received on the sale of said bonds, and he shall apportion thereto from year to year, in addition, amounts sufficient with the accumulations to extinguish at maturity the debt incurred by the issue of said bonds. The amount necessary to meet the annual sinking-fund requirements and to pay the interest on said bonds shall be raised by taxation from year to year.

SECTION 9. Sections six, seven, eight, nine, eleven, and thirteen of chapter four hundred and seventy-six of the acts of the year eighteen hundred and ninety-three are hereby repealed.

SECTION 10. This act shall take effect upon its passage.

[*Approved June 20, 1894.*]

ACTS OF 1895, CHAPTER 92

AN ACT MAKING APPROPRIATIONS FOR EXPENSES OF THE MASSACHUSETTS HIGHWAY COMMISSION

Be it enacted, etc., as follows:

SECTION 1. The sums hereinafter mentioned are appropriated to be paid out of the State Highway Loan fund, to meet expenses of the Massachusetts Highway Commission for the year ending on the thirty-first day of December in the year eighteen hundred and ninety-five, to wit:

For rent of office, including care, heating and lighting the same, a sum not exceeding one thousand dollars, this amount being in addition to the sum heretofore appropriated for rent in an act passed the present year.

For the salaries of clerks and such clerical assistance as said commission may find necessary, a sum not exceeding five thousand dollars.

For the salary of the chief engineer, a sum not exceeding three thousand dollars.

For incidental and contingent expenses of said commission, a sum not exceeding fifteen hundred dollars.

For traveling expenses of said commission, a sum not exceeding fifteen hundred dollars.

For expenses in connection with surveys of roads for the purpose of laying out and building State highways, a sum not exceeding ten thousand dollars.

SECTION 2. This act shall take effect upon its passage.

[*Approved March 7, 1895.*]

ACTS OF 1895, CHAPTER 347

AN ACT RELATIVE TO THE CONSTRUCTION OF STATE HIGHWAYS

Be it enacted, etc., as follows :

SECTION 1. The Massachusetts Highway Commission is hereby authorized to expend a sum not exceeding four hundred thousand dollars for the construction of State highways during the current year, in accordance with the provisions of chapter four hundred and seventy-six of the acts of the year eighteen hundred and ninety-three, and chapter four hundred and ninety-seven of the acts of the year eighteen hundred and ninety-four.

SECTION 2. No persons except citizens of this commonwealth shall be employed on the work authorized by this act.

SECTION 3. For the purpose of meeting any expenses which may be incurred under the provisions of this act, the treasurer and receiver-general is hereby authorized, with the approval of the governor and council, to issue scrip or certificates of indebtedness to an amount not exceeding four hundred thousand dollars, for a term not exceeding thirty years. Said scrip or certificates of indebtedness shall be issued as registered bonds or with interest coupons attached, and shall bear interest not exceeding four per cent. per annum, payable semi-annually on the first days of April and October in each year. Such scrip or certificates of indebtedness shall be designated on their face as the State Highway Loan, shall be countersigned by the governor, and shall be deemed the pledge of the faith and credit of the commonwealth, and the principal and interest thereof shall be paid at the times specified therein in gold coin of the United States or its equivalent; and said scrip or certificates of indebtedness shall be sold and disposed of at public auction, or in such other manner, at such times and prices, in such amounts, and at such rates of interest, not exceeding the rate above specified, as shall be deemed best. The sinking-fund established by chapter four hundred and ninety-seven of the acts of the year eighteen hundred and ninety-four shall also be maintained for the purpose of extinguishing bonds issued under the authority of this act, and the treasurer and receiver-general shall apportion thereto from year to year an amount sufficient with the accumulations of said fund to extinguish at maturity the debt incurred by the issue of said bonds. The amount necessary to meet the annual sinking-fund requirements and to pay the interest on said bonds shall be raised by taxation from year to year.

SECTION 4. This act shall take effect upon its passage.

[*Approved May* 1, 1895.]

ACTS OF 1895, CHAPTER 486

AN ACT RELATIVE TO THE CONSTRUCTION OF MACADAMIZED ROADS IN TOWNS

Be it enacted, etc., as follows :

SECTION 1. When a town of not less than ten thousand inhabitants, or not less than two nor more than five adjoining towns whose combined population does not exceed twelve thousand, vote at a town meeting to expend not less than three thousand dollars per year each year for the term of five years in the case of a single town, or four thousand dollars each year for the term of five years when not less than two nor more than five towns unite together, for macadamized roads, the commonwealth shall furnish out of the State Highway Loan authorized by chapter three hundred and forty-seven of the acts of the present year, through the Massachusetts Highway Commission, to such town or towns, free of charge, a steam road-roller of approved pattern and suitable size, for the sole use of such town or towns during said five years and as long thereafter as they continue to expend not less than fifty per cent. of the above-mentioned sum on macadamized roads each year; provided, nevertheless, that if said town or towns fail to expend said sum for macadamized roads in any one year, said road-roller shall then revert to the commonwealth. Said town or towns shall keep said roller in good repair.

SECTION 2. When not less than two nor more than five towns use a roller jointly, the town voting the largest proportion of the required sum shall have the first chance as to the time of using it, and may retain possession of it each year for a length of time proportionate to the sum voted by said town. The six months between the first day of May and the first day of November in each year shall be deemed the proper period for macadamizing roads.

SECTION 3. The Massachusetts Highway Commission shall not expend more than nine thousand dollars in carrying out the provisions of this act during the year eighteen hundred and ninety-five.

SECTION 4. This act shall take effect upon its passage.

[*Approved June 5, 1895.*]

RESOLVES OF 1896, CHAPTER 33

RESOLVE TO PROVIDE FOR PRINTING EXTRA COPIES OF THE REPORT OF THE MASSACHUSETTS HIGHWAY COMMISSION

Resolved, That three thousand extra copies of the third annual report of the Massachusetts Highway Commission be printed and bound in cloth. Out of the number so printed each member of the present General Court shall be entitled to receive ten copies, and the residue shall be distributed under the direction of the commission.

[*Approved March 25, 1896.*]

RESOLVES OF 1896, CHAPTER 86

RESOLVE RELATIVE TO A STATE HIGHWAY BETWEEN THE CITY OF BOSTON AND THE CITY OF NEWBURYPORT

Resolved, That the Massachusetts Highway Commission consider the expediency of laying out a State highway between the city of Boston and the city of Salem or the city of Newburyport, over the shore route, so called, which route may be described substantially as follows:

Starting from the south ferry, at Lewis street, in Boston, thence through Lewis street to Maverick Square, thence through Maverick Square to Chelsea street, thence over Chelsea street to Bennington street, thence over Bennington street to Orient Heights, thence over the main traveled road to the town of Revere, continuing on the main road to Beachmont, continuing over the main traveled road, known as Ocean Avenue, along the ocean front to the Point of Pines, crossing the Saugus River on the easterly side of the Boston, Revere Beach, and Lynn Railroad, and running to the south end of Sea street in Lynn, thence through Sea street to Broad street, thence through Broad street to Lewis street, thence through Lewis street to

New Ocean street, thence through New Ocean street to the town of
Swampscott, thence through New Ocean street, in Swampscott, to
the junction of Burrill street and Paradise road, thence over Paradise
road to the northeast end of said road, thence through Paradise
Woods on nearly a straight line to Vinin Square, at the junction of
the towns of Swampscott and Marblehead and the city of Salem,
thence northerly to Loring Avenue in the city of Salem, thence over
Loring Avenue to Lafayette street, thence over Lafayette street to
Central street, thence over Central street to Essex street, thence
through Salem to and over Beverly Bridge, thence through the city
of Beverly, and thence to Newburyport, using the present traveled
roads as far as may be, with such additions of new road as may be
necessary. Said Massachusetts Highway Commission shall report to
the next General Court the probable cost of such a highway, with such
other information as may be obtained in relation thereto, on or before the thirty-first day of January in the year eighteen hundred and
ninety-seven.

[Approved April 28, 1896.]

ACTS OF 1896, CHAPTER 345

AN ACT RELATIVE TO STATE HIGHWAYS

Be it enacted, etc., as follows :

SECTION 1. When a highway is laid out as a State road, the Massachusetts Highway Commission shall construct and maintain that
portion of the way between the inside lines of sidewalks upon either
side. The sidewalks of said road may be constructed and maintained
in accordance with the public statutes and amendments thereto, and
the provisions of section six of chapter four hundred and ninety-seven
of the acts of the year eighteen hundred and ninety-four shall only
apply to that portion of the way between the inside lines of sidewalks.
The inside lines of sidewalks referred to in this section are those
lines which are nearest to the center of the highway.

SECTION 2. A city or town in which a State road lies shall, at its
own expense, keep such road sufficiently clear of snow and ice so
that the same shall be reasonably safe for travel, as now required by
the public statutes and amendments thereto.

SECTION 3. Instead of filing the original petition with the county
commissioners, as now required by section two of chapter four hun-

dred and ninety-seven of the acts of the year eighteen hundred and ninety-four, it shall hereafter be sufficient to file a certified copy thereof with said county commissioners.

SECTION 4. This act shall take effect upon its passage.

[*Approved April 28, 1896.*]

ACTS OF 1896, CHAPTER 481

AN ACT RELATIVE TO THE CONSTRUCTION OF STATE HIGHWAYS

Be it enacted, etc., as follows :

SECTION 1. The Massachusetts Highway Commission is hereby authorized to expend a sum not exceeding six hundred thousand dollars for the construction of State highways, in accordance with the provisions of chapter four hundred and seventy-six of the acts of the year eighteen hundred and ninety-three, and chapter four hundred and ninety-seven of the acts of the year eighteen hundred and ninety-four.

SECTION 2. No persons except citizens of this commonwealth shall be employed on the work authorized by this act.

SECTION 3. For the purpose of meeting any expenses which may be incurred under the provisions of this act, the treasurer and receiver-general is hereby authorized, with the approval of the governor and council, to issue scrip or certificates of indebtedness to an amount not exceeding six hundred thousand dollars, for a term not exceeding thirty years. Said scrip or certificates of indebtedness shall be issued as registered bonds or with interest coupons attached, and shall bear interest not exceeding four per cent. per annum, payable semi-annually on the first day of April and of October in each year. Such scrip or certificates of indebtedness shall be designated on their face as the State Highway Loan, shall be countersigned by the governor, and shall be deemed a pledge of the faith and credit of the commonwealth, and the principal and interest thereof shall be paid at the times specified therein in gold coin of the United States or its equivalent; and said scrip or certificates of indebtedness shall be sold and disposed of at public auction, or in such other manner, at such times and prices, in such amounts, and at such rates of interest, not exceeding the rate above specified, as shall be deemed best. The sinking-fund established by chapter four hundred and ninety-seven of the acts of the year eighteen hundred and ninety-four shall also be maintained for the purpose of extinguishing bonds issued under the authority of this act, and the treasurer and receiver-general shall apportion thereto from year to year an amount sufficient with the accumulations of said fund to extinguish at maturity the debt incurred by the issue of said bonds. The amount necessary to meet the annual sinking-fund requirements and to pay the interest on said bonds shall be raised by taxation from year to year.

SECTION 4. This act shall take effect upon its passage.

[*Approved June 4, 1896.*]

ACTS OF 1896, CHAPTER 513

AN ACT TO PROVIDE FOR AIDING TOWNS IN THE CONSTRUCTION AND MAINTENANCE OF BETTER ROADS

Be it enacted, etc., as follows:

SECTION 1. Upon the application to the Massachusetts Highway Commission of the county commissioners of any county, made at the request of any town of not more than twelve thousand inhabitants within said county, there shall be furnished by said Highway Commission to said county, at the expense of the commonwealth, one or more steam-rollers, portable stone-crushers, and such other road machines as the said Highway Commission may deem necessary for the construction and maintenance of better roads in the town making such request. Such machines shall remain the property of the commonwealth and shall be managed and maintained under the direction of the county commissioners. The county commissioners shall engage competent engineers and skilled mechanics to operate said machines, who shall be paid from the county treasury such sums for each day's actual services as the county commissioners may determine. The expenses so incurred shall be repaid to the county by the towns using said machines, as apportioned by the county commissioners, in proportion to the time in which such machines were used by them.

SECTION 2. Chapter four hundred and eighty-six of the acts of the year eighteen hundred and ninety-five is hereby repealed.

SECTION 3. This act shall take effect upon its passage.

[*Approved June 6, 1896.*]

ACTS OF 1896, CHAPTER 548

AN ACT MAKING APPROPRIATION FOR EXPENSES AUTHORIZED BY THE PRESENT LEGISLATURE AND FOR CERTAIN OTHER EXPENSES AUTHORIZED BY LAW

Be it enacted, etc., as follows:

For expenses in connection with aiding towns in the construction and maintenance of better roads, as authorized by chapter five hundred and thirteen of the acts of the present year, a sum not exceeding twelve thousand dollars.

ACTS OF 1896, CHAPTER 541

AN ACT RELATIVE TO STREET RAILWAYS LOCATED ON STATE HIGHWAYS

Be it enacted, etc., as follows :

SECTION 1. Whenever in the construction of a State highway it becomes necessary, in the opinion of the Massachusetts Highway Commission, to change the location, relay or change the grade of that part of any street-railway located on said highway, or to place different material between its tracks, or to make any other change in the location and construction of said railway, said commission may, in the manner provided in section twenty-two of chapter one hundred and thirteen of the public statutes for making such changes by boards of aldermen and selectmen, order the company owning or operating said railway to make such changes; provided, however, that the company shall thereafter enjoy the same rights in the new location that it had in the original location; and, unless the same are made within the time limited by said commission, the commission may make said changes, and the cost of making the same, whether by the railway company or by said commission, shall be paid by said commission; said cost with interest at a rate not exceeding four per cent. per annum shall be paid by said railway company to the commonwealth in ten equal annual payments; and the auditor of the commonwealth on or before the first day of July in each year shall certify the amount due to the tax commissioner, who shall forthwith demand the same; and payment shall be made within thirty days thereafter. The claim of the commonwealth shall have priority over all other claims against

17

said railway company, except for labor, and shall be collected in the same manner as the corporation tax; but any such company may itself pay for the expenses of said changes at the time of making the same, and may anticipate said annual payments in whole or in part.

SECTION 2. This act shall take effect upon its passage.

[*Approved June* 9, 1896.]

APPENDIX B

SHOWING LABORATORY EXPERIMENTS ON ROAD-BUILDING STONES [1]

THE following described results were obtained in the highway laboratory of the engineering department of the Lawrence Scientific School of Harvard University. Those under the head "Coefficient of Abrasion" were obtained by the Deval method, which has been employed for some time by the French engineers for determining the relative value of the stone used in the construction and maintenance of the national highways of France. These results are said to agree well with those obtained in actual practice.

The apparatus used in the tests consists of a cast-iron cylinder 20 cm. in diameter and 34 cm. in depth. At one end is an opening which can be closed with a tightly fitting iron cover. This cylinder is mounted on an axle at an angle of 30° with the axis of the cylinder, and is supported on an iron frame. At one end of the axle is a pulley-wheel by which the cylinder is revolved; at the other is an instrument which records its revolution.

The stone to be tested is first broken into pieces between 6.31 cm. and 3.18 cm. in diameter, which are carefully washed to remove any foreign matter. In the cylinder are placed 5 kilograms of this stone. The top is then bolted on, and the cylinder is made to revolve for five hours at the rate of 2000 revolutions an hour, making in all 10,000 revolutions. By this process the stones are thrown from one end of the cylinder to the other, and at the same time are rolled against the sides of the vessel and against one another. When 10,-

[1] Reprinted from annual report of Massachusetts Highway Commission for 1896.

000 revolutions are completed the cover is removed and the contents emptied into a tray. The cylinder is then thoroughly washed to remove the dust that adheres to its sides. Each stone above 3.18 cm. in diameter is then washed under the same water. This water is then filtered, and the filtrate when dry is mixed with the detritus taken from the cylinder. The detritus is then put into a sieve, by which it is separated automatically into seven sizes. These seven sizes, together with the stones that have not been worn below 3.18 cm. in diameter, are each carefully weighed and their weights recorded.

The amount of detrition under .16 cm. is rarely less than 20 grams per kilogram of stone used; therefore 20 has been adopted as the standard, and the coefficient of quality is obtained by the following formula :

$$q = 20 \times \frac{20}{u} = \frac{400}{u},$$

in which u represents the weight in grams of detritus per kilogram of stone.

It seemed well, in beginning this work, to be guided as far as possible by the experience of others ; and for this reason the Deval test was adopted, for it appeared to be the only practicable method of testing road metals yet devised. After a number of trials were completed with the Deval apparatus, and their results studied, it was recognized that all the valuable properties possessed by a good road metal were not embraced in this test. The value of any good stone as a road metal is due to certain properties possessed by it. Among these there are three which stand prominent—cementing value, toughness, and hardness. It is evident that the Deval apparatus does not test the very important property of cementing value in the different road metals. The commission, recognizing this deficiency, accordingly directed its attention to devising some means of supplying it. As no previous attempt has been made in this direction, the commission had to invent its own method, which is as follows :

The stone to be tested is ground to a powder and passed through a sieve of .25 mm. The powder is then put in a slightly tapered steel die of circular section, about 3 cm. diameter, mixed with water, and subjected to a pressure of 2300 kilograms. The resulting briquet is then put aside for at least one week, so it may thoroughly dry.

It was at first thought that a test by direct compression would determine the cementing power of the stone. A number of briquets were tried in this way, but the results were not very satisfactory. On further consideration it appeared that a test by impact would more thoroughly determine the cementing power of the stone than that by compression, and this method would have the further advantage of approximating more closely to the actual conditions obtaining on roads ; accordingly a machine was devised for testing the briquets by impact. With this machine a hammer one kilogram in weight can be dropped freely from any desired height upon a plunger under which the briquet to be tested is placed. The hammer works automatically, and is tripped at the desired height. Attached to the plunger is a lever pivoted at one sixth of its length from the plunger, and carrying a pencil at its free end. The pencil has a vertical movement five times as great as that of the plunger, and its movement is registered on a drum against which the pencil presses. The drum rotates through a small angle at each stroke of the hammer. An automatic diagram is thus taken of the behavior of the briquet throughout the whole test.

An analysis of the diagram so taken shows at once the number of blows required to cause the destruction of the briquet. A very interesting point is brought out by these diagrams, viz. : in every case the diagram shows that the plunger rebounded at each stroke until the briquet began to fail. This behavior is exactly analogous to the elastic phenomena observed in all materials of construction ; consequently the point at which the briquet ceases to rebound corresponds to the elastic limit of the material. Beyond this point the briquet falls to pieces rapidly.

Briquets were made from many kinds of stone and were tested in this machine. It was thought desirable to use a constant blow for all the briquets, and a short experience indicated a fall of 3 cm. as suitable, since it broke the most tenacious materials with a moderate number of blows, and yet was not too great to permit the careful determination of the properties of the poorer stones. All the briquets were 2.5 cm. high.

The surface of a macadamized road is constantly being abraded and recemented. Evidently a road made from a material which has the property of recementing in a high degree will keep in better condition than one made from a material of lower recementing power. It was therefore desirable to determine the recementing properties of the stones tested. A new set of briquets was made, differing from the former only in that they were of constant weight instead of con-

18

stant height.   These were tested in the manner described above, and
then were remade and retested.

It has not been thought desirable to present herewith the complete
data obtained from the impact test, as the series is not yet completed.
The writer has, however, collected and shown in accompanying table
some of the more important results thus far obtained, a sufficient
number to indicate the scope of the work done.   In this table the
stones are arranged in the order of their power of resisting abra-

TABLE SHOWING SPECIFIC DENSITIES, COEFFICIENTS
ING VALUES OF

		City or Town.	1. Specific Density.
31,	VII.	(16-V), Saugus, Essex Co., Mass.	3.03
31,	II.	(32-L), Newton, Middlesex Co., Mass.	2.80
36,	VI.	(32-G), Newbury, Essex Co., Mass.	....
38,	I.	(16-D), Lynn, Essex Co., Mass.	3.03
38,	I.	(12-C), Lynn, Essex Co., Mass.	3.03
31,	VII.	(20-V), Saugus, Essex Co., Mass.	3.03
38,	I.	(10-C), Lynn, Essex Co., Mass.	2.99
36,	VI.	(7-P), Salisbury, Essex Co., Mass.	....
36,	VI.	(3-H), Newburyport, Essex Co., Mass.	....
31,	VII.	(21-Y), Saugus, Essex Co., Mass.	3.01
31,	VI.	(25-P), Boston, Suffolk Co., Mass.	....
31,	VII.	(24-W), Saugus, Essex Co., Mass.	3.01
31,	IV.	(31-N), Medford, Middlesex Co., Mass.	3.03
12,	II.	(25-R), West Springfield, Hampden Co., Mass.	2.96
37,	VI.	(31-F), Salem, Essex Co., Mass.	2.92
38,	III.	(30-K), Quincy, Norfolk Co., Mass.	2.96
31,	VI.	(7-B), Brookline, Norfolk Co., Mass.	2.99
38,	I.	(12-K), Lynn, Essex Co., Mass.	2.66
36,	VI.	(29-I), Newbury, Essex Co., Mass.	....
31,	VIII.	(6-N), Everett, Middlesex Co., Mass.	2.87
44,	III.	(26-L), Duxbury, Plymouth Co., Mass.	2.68
31,	VII.	(26-T), Revere, Suffolk Co., Mass.	2.65
43,	II.	(32-F), Gloucester, Essex Co., Mass.	....
43,	IV.	(34-K), Rockport, Essex Co., Mass.	....
		Meriden, Conn.	2.83
38,	I.	(12-N), Lynn, Essex Co., Mass.	....
		Chester, Hampden Co., Mass.	....
26,	VIII.	(23-V), Waltham, Middlesex Co., Mass.	2.62
		Lee, Berkshire Co., Mass.	....
		Lee, Berkshire Co., Mass.	2.60
5,	III.	(16-K), Lee, Berkshire Co., Mass.	
31,	VI.	(5-B), Brookline, Norfolk Co., Mass.	2.87
43,	II.	(28-U), Gloucester, Essex Co., Mass.	2.64
		Northampton, Mass.	2.74
		Chester, Hampden Co., Mass.	....
32,	VII.	(3-P), Quincy, Norfolk Co., Mass.	2.66
		Plymouth, Plymouth Co., Mass.	....
13,	IX.	(1-R), Orange, Franklin Co., Mass.	....

sion. Column 1 contains the specific density of the stones; column 2 the coefficients of abrasion (determined in the manner previously described); the next column gives the number of blows required to stress the 2.5-cm. briquets to their elastic limits; column 4 gives the same data for the first testing of the 30-gram briquets prepared for the recementation test, and the next column gives the number of blows that the recemented briquets will stand before reaching their elastic limits.

OF ABRASION, CEMENTING VALUES, AND RECEMENT-STONES TESTED

2. Coefficient of Wear.	3. Cementing Value.	4. Cementing Value of 30-Gram Briquet.	5. Recementing Value of 30-Gram Briquet.	NAME OF STONE.
21.22	....	....	....	Diabase.
20.79	....	....	....	Trachyte.
20.40	....	....	....	Olivin diabase.
20.37	....	56	29	Diabase.
19.77	....	....	....	Diabase.
18.25	....	....	....	Diabase.
18.17	....	....	....	Diabase.
16.76	....	....	....	Camptonite.
16.10	....	40	34	Diabase.
16.08	....	42	20	Diabase.
16.08	23	109	31	Felsite.
16.02	....	121	39	Diabase.
15.82	....	....	....	Diabase.
15.60	82	....	....	Olivin diabase (poor specimen).
15.55	27	58	19	Augite diorite (poor specimen).
15.21	....	....	....	Diabase.
14.71	....	81	28	Olivin diabase.
14.66	....	....	....	Felsite.
14.45	....	....	....	Quartz diorite.
13.87	....	....	....	Olivin diabase (coarse-grained).
13.46	....	....	....	Hornblende granitite.
13.21	....	11	9	Felsite.
12.63	5	....	....	Augite nephelin syenite.
12.57	....	....	....	Hornblende granitite.
12.50	....	....	....	Augite diabase (poor specimen).
12.30	6	....	....	Felsite.
12.21	11	....	....	Schist.
12.16	16	....	....	Hornblende granitite.
12.15	....	....	....	Schist.
11.65	....	....	....	Quartzite.
11.43	23	....	....	Gneiss.
11.40	31	....	....	Hornblende diorite.
11.03	6	....	....	Hornblende granitite.
10.69	....	....	....	Hornblende granitite.
10.62	35	....	....	Magnetite corundum gneiss.
10.16	....	....	....	Hornblende granitite.
10.10	11	18	12	Field stone (erratics).
9.78	12	27	17	Hornblende granitite.

TABLE SHOWING SPECIFIC DENSITIES, COEFFICIENTS OF
VALUES OF STONES

	CITY OR TOWN.	1. Specific Density
3, IV. (23-T),	Great Barrington, Berkshire Co., Mass...............	....
	Nantucket, Nantucket Co., Mass.	....
2, VII. (28-D),	Pittsfield, Berkshire Co., Mass......................	2.82
31, V. (14-T),	Somerville, Middlesex Co., Mass......................	2.86
	North Attleborough, Mass.	....
	Diamond Hill, Cumberland, R. I......................	....
31, VII. (12-W),	Saugus, Essex Co., Mass.	....
	Tisbury, Dukes Co., Mass.	....
	Mattapoisett, Mass................................	....
31, III. (1-V),	Newton, Middlesex Co., Mass.	....
31, V. (7-M),	Somerville, Middlesex Co., Mass....................	2.75
2, VIII. (24-R),	Lenox, Berkshire Co., Mass........................	2.84
10, II. (26-B),	Buckland, Franklin Co., Mass......................	....
	Tomkins Cove, N. Y...............................	2.84
	Holden, Mass.	....
	Whitman, Mass....................................	....
	Paxton, Worcester Co., Mass.......................	....
2, IX. (15-Z),	Lee, Berkshire Co., Mass.	2.74

An inspection of the above table suggests some general considerations as
to the relative values for road materials of different varieties of stone, but it
appears to the writer to be injudicious, with the limited data at hand, to discuss
them.

ABRASION, CEMENTING VALUES, AND RECEMENTING
TESTED (*Concluded*)

2. Coefficient of Wear	3. Cementing Value.	4. Cementing Value of 30-Gram Briquet.	5. Recementing Value of 30-Gram Briquet.	NAME OF STONE.
9.52	....	....	....	Limestone.
9.47	....	....	....	Field stone (erratics).
9.38	15	....	....	Limestone.
9.28	34	....	....	Diabase (very coarse-grained).
9.09	....	....	....	Field stone (erratics).
9.07	9	....	....	Quartzite.
8.99	....	....	....	Hornblende granitite.
8.88	9	....	....	Field stone (erratics).
8.78	....	....	....	Field stone (erratics).
8.67	14	....	....	Conglomerate.
8.48	21	....	....	Slate (Cambrian?).
8.04	27	....	....	Schist.
7.94	8	....	....	Hornblende gneiss.
7.84	16	....	....	Limestone (silicious).
6.58	....	....	....	Field stone (erratics).
5.93	....	....	....	Field stone (erratics).
5.01	....	....	....	Granitoid gneiss.
2.85	....	....	....	Marble.

In closing, it may be well to state that methods for testing respectively the toughness and hardness of road metals have also been devised, but they have not been employed a sufficient time to give enough results to warrant their tabulation.

LOGAN WALLER PAGE.

18*

Town.	Contractor.	Excavation.						Lodge Excavation.
		All Kinds.	Surface Grading.	Sand.	Gravel.	Hard-pan.	Clay.	
		Cubic Yard.	Cubic Yard.	Cubic Yard.	Cubic Yard.	Cubic Yard.	Cubic Yard.	Cubic Yard.
Andover	Town	$ 0 25						$ 1 50
Ashby, 1894	Town	25	$ 0 30					1 75
Ashby, 1895	Town	30						1 75
Athol	Town	30						1 75
Auburn	Town	40						1 75
Beverly	City	45						1 75
Brewster	Town			$ 0 25	$ 0 25	$ 0 25	$ 0 40	1 75
Buckland, 1894	Town	30						1 75
Buckland, 1895	Town	35						1 75
Cottage City, 1894	Town	25						......
Cottage City, 1895	Town	25						......
Dalton	Town	35						1 75
Deerfield	Town			30	30	30	50	1 75
Dennis	Town	25						1 75
Duxbury, 1894	Town	25						1 75
Duxbury, 1895	Town	25						1 75
Easthampton	Town	35						1 75
Fairhaven, 1894	Town	27						......
Fairhaven, 1895	Town	35						1 75
Fitchburg, 1894	City	25						......
Fitchburg, 1895	City	30	40					1 75
Gloucester, 1894	City	50⁴ / 30	40					1 75
Goshen, 1894	Town	30						1 75
Goshen, 1895	Town	40						1 75
Granby	Town	30	30					1 75
Great Barrington	Town	25	35					1 75
Hadley, 1894	Town	30						......
Hadley, 1895	Town	30						......
Hancock	Hendrick, Taylor & Warner	30						2 00

1 Red granite; not local.   2 Local trap.
4 Sand.   5 Opening three feet or less.

DIX C

STATE ROADS DURING 1894-95

TRACT PRICES

Rubble Masonry.			Telford.	Shaping.	Broken Stone.		Vitrified Clay Pipe, 12-inch.	Fencing (Guard-rail).	Side Drains.	Cobble Gutter.	Stone Bounds.
Dry.	Cement.	Gravel.			Local.	Trap.					
Cubic Yard.	Cubic Yard.	Cubic Yard.	Sq. Yard.	Sq. Yard.	Ton.	Ton.	Linear Foot.	Linear Foot.	Linear Foot.	Sq. Yard.	Each.
$2 50	$4 00	$0 40	$0 33	$0 02	$1 32	$2 05	$0 60	$0 15	.....	$0 60	$1 25
2 50	.....	30	20	.....	1 67	.....	60	10	$0 25	60	1 25
2 50	4 50	50	25	02	1 17	.....	60	15	25	50	1 25
2 50	4 50	40	33	02	.....	{ 1 98½ / 2 18 }	60	15	33	50	1 50
3 00	5 00	40	33	02	.....	2 30	60	15	33	50	1 50
2 50	5 00	50	35	02	.....	2 00²	60	15	35	60	1 25
3 00	5 00	.....	30	02	1 83	.....	60	15	33	50	1 25
2 50	.....	.....	.....	.....	.....	1 92½	.....	12	.....	65	1 00
2 50	5 00	50	33	02	.....	{ 1 85½ / 2 05 }	60	15	33	50	1 50
.....	.....	.....	.....	.....	1 97	.....					
.....	5 00	.....	.....	02	1 91³	.....	60	16	.....	.....	1 50
2 50	5 00	60	33	02	.....	1 95	60	15	33	60	1 50
2 50	5 00	60	33	02	.....	2 10	60	15	33	65	1 50
2 50	4 00	50	33	02	1 53	.....	60	15	30	65	1 25
2 50	.....	50	33	.....	1 47	.....	60	10	25	70	1 00
2 50	4 00	50	33	02	1 52	.....	60	15	25	70	1 00
3 00	5 00	50	30	02	.....	1 68	60	15	33	50	1 50
2 50	.....	.....	.....	.....	1 40	.....	60	12	25	70	1 00
2 50	5 00	70	33	02	1 45	.....	60	15	33	50	1 50
2 00	.....	50	25	.....	.....	1 83	60	10	25	.....	1 00
2 00	2 50	50	30	02	.....	2 00	60	15	.....	50	1 00
2 50	.....	60	33	.....	1 24	.....	60	12	25	65	1 50
{ 2 50⁶ / 4 00⁷ }	.....	50	25	.....	1 53	.....	65	{ 10⁶ / 12⁹ }	25	75	1 25
2 50	4 50	50	35	02	1 75	.....	60	15	35	60	1 50
2 50	.....	30	33	.....	.....	2 85	60	12	35	.....	1 00
2 50	.....	60	33	.....	.....	2 20	60	12	25	70	1 25
.....	.....	50	33	.....	.....	1 93	.....	12	25	.....	75
3 00	5 00	40	.....	02	.....	2 00	60	15	.....	.....	1 25
4 00	.....	.....	.....	.....	.....	.....	53	15	19	.....	50

² Also beach stone for foundation, ninety cents per ton.
⁶ Unpainted.   ⁷ Opening more than three feet.   ⁹ Painted.

TABLE OF CONTRACT

Town.	Contractor.	All Kinds.	Surface Grading.	Sand.	Gravel.	Hard-pan.	Clay.	Ledge Excavation.
		Cubic Yard.	Cubic Yard.	Cubic Yard.	Cubic Yard.	Cubic Yard.	Cubic Yard.	Cubic Yard.
Hingham ¹	Town							
Holbrook-Weymouth,	W. T. Davis	$0 43						$1 50
Holden, 1894	Town	35	$0 50					1 75
Holden, 1895	Town	35						1 75
Huntington	Town	30						1 50
Lee, 1894	Town	30	30					1 50
Lee, 1895	Town	35						1 75
Leicester, 1894	Town	⁵					$0 35	1 75
Leicester, 1895	Town			$0 25	$0 25	$0 35	35	1 75
Lexington	Town			35	35	50	50	1 75
Lincoln	Town	35						1 75
Marion, 1894	Town	25						
Marion, 1895	Town			35	35	60		1 75
Marshfield	Town	25						
Mattapoisett, 1894	Town	25						
Mattapoisett, 1895	Town	35						1 75
Middleborough, 1894	Town	30						1 75
Middleborough, 1895	Town	25						1 75
Monson	Town	30						1 75
Nantucket, 1894	Town	20						
Nantucket (1), 1895	Town	20						
Nantucket (2), 1895	Town	20						
Norfolk	Hendrick, Taylor & Warner	28						
North Adams	Town	25	30					1 75
Northampton	City	30						
North Attleborough, 1894	Town	30	30					
North Attleborough, 1895	Town	35						1 75
Norwood	Town			35	35	50	50	1 75
Orange, 1894 and 1895	Town	25						1 75
Paxton (1)	Town	35	40					1 75
Paxton (2)	Town	35						1 75
Pittsfield	City	25	35					1 50
Plymouth, 1894	Town	25	25					1 75
Plymouth, 1895	Town	25						1 75
Rehoboth	Town			35	35	35	35	1 75
Russell, H., 1894	Town	30	30					1 50
Russell, F., 1894	Town	30	30					1 50
Russell, H., 1895	Town	30						1 75
Scituate	Town	35						1 75
Shelburne, 1894	Town	30						1 75
Shelburne, 1895	Town	25						1 75
Shrewsbury	Town	40						1 75
Somerset	Town			25	30	60	60	1 75
Taunton	City	40						1 75

¹ Opening three feet or less.   ² Opening more than three feet.
⁵ Also common wall, $2.50 per rod; face wall, $3.50 per rod.
⁹ Not trap; granite from quarry.

PRICES (Continued)

| Rubble Masonry. | | Gravel | Telford | Shaping | Broken Stone. | | Vitrified Clay Pipe, 12-inch. | Fencing (Guard-rail). | Side Drains. | Cobble Gutters. | Stone Bounds. |
Dry.	Cement.				Local.	Trap.					
Cubic Yard.	Cubic Yard.	Cubic Yard.	Sq. Yard.	Sq. Yard.	Ton.	Ton.	Linear Foot.	Linear Foot.	Linear Foot.	Sq. Yard.	Each.
$3 50	......						$0 50	$0 15	......	......	$1 00
2 50	......	$0 50	$0 35		$1 57				$0 25	$0 50	1 00
2 50	$5 00	{ 50⁴ / 65 }	33	$0 02	1 56		60	15	33	50	1 50
2 50	4 50	40	25	02		$1 84	60	14	33	50	1 00
{ 2 50¹ / 4 50² }	{ 4 50 }	65	35		1 70	2 20	60	12	25		1 25
3 50	5 00	45	33	02		2 40	60	15	35	60	1 35
2 50		1 00	25	05	1 37	2 10		12	25	60	......
2 50	4 00	65	30	02	1 43	2 10	60		25	60	1 50
2 50⁶	5 00	60	33	02	1 55		60	15	35	60	1 50
2 50	4 50	40	30	02	1 40		60	15	33	50	1 25
2 50		50			1 45		60	12			75
2 50	5 00	75	30	02	{ 1 45 / 1 25⁷ }	{ .... }	60	15	33	60	1 50
2 50		50	33		1 47		60	10	25	70	1 00
2 50			33		1 30		60	12	25		1 00
2 50	5 00	50	30	02	1 25		60	15	33	50	1 50
2 50		50	33		1 47		60	12	25	65	1 25
2 50	4 00	50	33	02	1 28		60	15	25	65	1 00
2 50	4 50	35	33	02	1 72		60	12	25	75	1 00
......		1 50				2 82⁸	......	......			75
......	5 00	1 75		02		3 00⁸	60	15		50	1 25
......	5 00	1 75		02		3 00⁸	60	15		50	1 25
2 50	6 00	......					60	20			1 00
5 00		60	33			1 92	60	12	25	70	1 25
......	......				1 67	.....					
......					1 06		55			70	......
2 50	5 00	50	30	02	1 37		60	15	33	.. ..	1 50
3 00	5 00	60	35	02	1 56		60	15	35	60	1 25
2 50	4 00	40	33	02	1 59	2 12	60	12	25	70	1 25
2 50		75	30		1 30		60	13	30	65	1 50
2 50	4 50	65	30	02	1 40		60	15	30	65	1 50
2 50	.....	50	33			1 88	60	12	25	......	1 00
2 50		50	33		1 82		60	12	25		1 25
2 50	5 00	50	.....	02	1 44		60	12	.....	60	1 25
2 50	5 00	55	30	02	1 48		60	15	33	60	1 50
2 50		40	25			2 00	......	12	27	75	1 50
2 50		40	25			2 00	......	12	27	75	1 50
2 50	4 50	40	25	02		2 00	60	14	33	50	1 50
2 50	4 50	40	30	02		1 85₉	60	15	33	50	1 25
2 50		......			1 42		60	12	......		1 25
2 50¹⁰	4 50	......	30	02	1 43		60	15	30	65	1 25
2 50	5 00	55	33	02	1 52		60	15	33	50	1 50
2 50	5 00	50	33	02	1 50		60	15	33	60	1 50
3 00	4 50	60	33	02	1 66		60	15	40	50	1 25

3 Lump sum; gravel road.   4 Different hauls.   5 Included in shaping.
7 Tailings.   8 Limestone from Tompkins Cove.
10 Also wall masonry, $3.50 per cubic yard.

| Town. | Contractor. | Excavation | | | | | | Ledge Excavation. |
| | | All Kinds. | Surface Grading. | Sand. | Gravel. | Hard-pan. | Clay. | |
		Cubic Yard.	Cubic Yard.	Cubic Yard.	Cubic Yard.	Cubic Yard.	Cubic Yard.	Cubic Yard.
Tisbury	Town	$0 20	$0 25					
Tyngsborough	Town							
Walpole	Town	30	30					
Watertown	Town	40						$1 75
Westfield	Town	25	30					
Westminster, 1894	Town	25	30					1 75
Westminster, 1895	Town	35						1 75
West Newbury	C. H. Kelleher	22						1 40
Westport	Town	25	35					1 75
West Springfield	Town	30						1 75
West Tisbury	Town	...²		$0 25	$0 30		$0 35	1 75
Weymouth	Town			35	35	$0 35	35	1 75
Whitman, 1894	Town	40	40					1 75
Whitman, 1895	Town			40	40	40	40	1 75
Wilbraham	M. R. Fisk	27						
Williamstown	Town	35						1 75
Wrentham, 1894	Town	30	30					1 75
Wrentham, 1895	Town	40						1 75
Yarmouth, N., 1894	Town	25	30					1 75
Yarmouth, N., 1895	Town	25						1 75
Yarmouth, S., 1895	Town	25						1 75
Averages		$0 301	$0 330	$0 313	$0 323	$0 420	$0 423	$1 725

1 Local trap. 2 Grubbing, $100. 3 Long haul,
5 Opening over six feet.

PRICES (*Concluded*)

Rubble Masonry.		Gravel	Telford	Shaping	Broken Stone.		Vitrified Clay Pipe, 12-inch	Fencing (Guard-rail).	Side Drains.	Cobble Gutters.	Stone Bounds.
Dry.	Cement.				Local.	Trap.					
Cubic Yard.	Cubic Yard.	Cubic Yard.	Sq. Yard.	Sq. Yard.	Ton.	Ton.	Linear Foot.	Linear Foot.	Linear Foot.	Sq. Yard.	Each.
$ 3 00	......	......	$ 0 25	......	$ 1 67	......	$ 0 65	$ 0 12	$ 0 25	......	$ 1 00
......	......	$ 0 45	......	$ 0 02	......	$ 2 13	......	15	......	......	1 25
2 50	......	50	33	......	1 82	......	60	12	25	$ 0 65	1 00
3 00	$ 5 00	50	33	02	......	1 92	60	15	33	50	1 50
2 50	......	50	25	......	......	1 50	60	10	25	......	1 00
2 50	5 00	50	33	......	......	1 92	60	10	25	......	1 10
2 50	· 5 00	50	33	02	......	1 95	60	15	30	60	1 10
2 70	3 25	60	65	02	......	1 35	65	19	30	85	1 00
2 50	......	47	33	......	1 41	......	60	12	25	......	1 25
3 00	6 00	40	35	02	......	1 52[1]	60	15	35	70	1 25
2 50	4 00	......	25	02	1 70	......	60	16	25	65	1 25
3 00	5 00	40	30	02	......	2 10[6]	60	15	33	50	1 50
2 50	......	40[2]	33	......	1 88	......	60	12	25	70	2 00
3 00	5 00	60	30	02	1 66	......	60	15	33	50	1 50
2 50	4 50	36	......	02	......	2 00	50	13	......	50	1 25
2 50	5 00	65	33	02	......	{ 1 90[4] / 2 10 }	60	15	33	50	1 50
2 50	......	50	25	......	1 48	......	60	12	25	......	1 00
2 75	4 50	60	33	02	1 70	......	60	15	30	55	1 25
2 50	......	50	......	......	1 38	......	60	12	......	75	1 00
3 00	4 50	1 00	......	02	1 59	......	60	15	......	60	1 25
..... {	{ 3 50 / 8 50[5]	} 60	......	02	......	2 11[6]	60	12	......	60	75
$ 2 67[6]	$ 4 75[0]	$ 0 56[7]	$ 0 31[5]	$ 0 02[0]	$ 1 51[8]	$ 2 05[9]	$ 0 59[8]	$ 0 13[7]	$ 0 29[2]	$ 0 60[1]	$ 1 25[6]

$0.50 per cubic yard.   4 Red granite from quarry.
          6 Not trap; granite from quarry.

APPENDIX D

LIST OF IMPORTANT WORKS ON HIGHWAY CONSTRUCTION

THE following titles are those of books which are deemed of value to the general reader. The list omits many valuable publications which are designed particularly for the use of engineers, and also some of a popular nature which in effect duplicate those given below :

BYRNE, AUSTIN T. A Treatise on Highway Construction : Designed as a Text-book and Work of Reference for all who may be Engaged in the Location, Construction, or Maintenance of Roads, Streets, and Pavements. New York, 1892. 8vo, pp. xxxiv., 686.

EGLESTON, NATHANIEL HILLYER. The Home and its Surroundings ; or, Villages and Village Life. With Hints for their Improvement. New and revised edition. New York, Harper & Brothers, 1884. Illus., pp. 326.

GILLESPIE, W. M. Manual of the Principles and Practice of Roadmaking: Comprising the Location, Construction, and Improvement of Roads and Railroads. New York, A. S. Barnes & Co., 1847. Illus., pp. 336.

HERSCHEL, CLEMENS. Prize Essay on Roads and Road-making. Boston, Wright & Potter, 1870. pp. 63.

JENKS, JEREMIAH WHIPPLE. Road Legislation for the American State. Baltimore, 1889. American Economic Association Publications, vol. iv., pp. 145–227.

LAW, HENRY. Rudiments of the Art of Constructing and Repairing Common Roads. To which is Prefixed a General Survey of the Principal Metropolitan Roads, by S. Hughes. Second edition, with additions. London, John Meade, 1855. Illus., pp. 158.

MACADAM, J. L. Remarks on the Present System of Road-making.